THE BEGINNINGS OF THE MODERN PHILOSOPHY OF MUSIC IN ENGLAND

To Alan
with esteem
From Jamie

In memory of
Peter T. Millard

The Beginnings of the Modern Philosophy of Music in England

Francis North's *A Philosophical Essay of Musick* (1677)
with comments of
Isaac Newton, Roger North
and in the
Philosophical Transactions

JAMIE C. KASSLER
Fellow, The Australian Academy of the Humanities

ASHGATE

Published by
Ashgate Publishing Limited
Gower House
Croft Road
Aldershot
Hampshire GU11 3HR
England

Ashgate Publishing Company
Suite 420
101 Cherry Street
Burlington, VT 05401-4405
USA

Ashgate website: http://www.ashgate.com

British Library Cataloguing in Publication Data
Kassler, Jamie Croy
The beginnings of the modern philosophy of music in England :
Francis North's A philosophical essay of musick (1677)
with comments of Isaac Newton, Roger North and in the
Philosophical Transactions
1. Guilford, Francis North, Baron, 1637-1685. Philosophical
essay of musick 2. Music – Philosophy and aesthetics –
History
I. Title
780.1

Library of Congress Cataloging-in-Publication Data
Kassler, Jamie Croy.
The beginnings of the modern philosophy of music in England :
Francis North's A philosophical essay of musick (1677) with comments of Isaac
Newton, Roger North and in the Philosophical Transactions.
p. cm.
Includes bibliographical references (p.) and indexes.
Contents: Introduction to A philosophical essay of musick (1677) – The
system of the Essay – The critics of the Essay – Appendix A, Editions –
Appendix B, Some "ingenious acquaintances" of Francis North.
ISBN 0-7546-0139-0 (alk. paper)
1. Music–England–Philosophy and aesthetics–History–17th century. 2.
Music–Acoustics and physics–History–17th century. 3. Guilford, Francis
North, Baron, 1637-1685. Philosophical essay of musick. 1. Guilford, Francis North, Baron,
1637-1685. Philosophical essay of musick. II. Title.

ML3800.K186 2004
781'.1–dc21

2002043898

ISBN 0 7546 0139 0

Typeset in Times New Roman by Computer Music Company, Sydney, Australia.
Printed and bound in Great Britain by MPG Books Ltd, Bodmin, Cornwall

CONTENTS

Illustrations vii
Prefatory Note and Acknowledgments ix
Abbreviations xi

1 Introduction to *A Philosophical Essay of Musick* (1677) 1
Hypothetical Physics 1650–76 5
The 1677 *Essay* 21

2 The System of the *Essay* 37
First Part: The Physical Hypothesis 40
Second Part: The Mechanics of Music 48
Third Part: The Semantics of Music 76

3 The Critics of the *Essay* 89
The 1677 Critics 91
The Later Critic 104

Conclusion 121

Appendix A: Editions 131
Editorial Note 131
1. Francis North, A Philosophical Essay of Musick (1677) 135
2. Notice of the *Essay* in the *Philosophical Transactions* (1677) 171
3. Isaac Newton, Letter to John North concerning the *Essay* (1677) 175
4. Roger North, 'Some Notes upon an Essay of Musick' (c.1694) 179

Appendix B: Some 'Ingenious Acquaintances' of Francis North 201

References 213
Index of Names 229
Index of Subjects 233

ILLUSTRATIONS

CHAPTER FIGURES

1.1 Charleton's 'stream lines' from (a) the sounding body to (b) the ear 6
1.2 Hobbes's demonstration of sound reception 12
1.3 Francis North, 1st baron Guilford (1683) 22
1.4 Title page: *A Philosophical Essay of Musick* 24

2.1 Flue pipe 59
2.2 Simple string pendulum 63
2.3 Non-rigid compound pendulum 64
2.4 Double pendulum 65
2.5 Trumpet marine 67

3.1 Wallis's representation of periodicities 90
3.2 Hooke's law of elasticity demonstrated 93
3.3 Flue pipe: Initiation of vibration according to (a) Francis North and (b) Roger North 114

APPENDIX A FIGURES

1.A *Graphs showing the periodicity patterns of consonances in relation to the fundamental tone as a function of time* 137
1.B *First six notes of the harmonic series arranged as a major chord over a fundamental bass* 138
1.C *A monochord* 159
1.D *Clarification of* Figure 1.C 167
3.A *Representation of sound reception from two tonal stimuli* 177
4.A *Barometer within a barometer* 180
4.B *Triangles:* (i) *isoceles,* (ii) *scalene* 189
4.C *Architectural order of three columns* 189
4.D *Flue pipe as driven by edge tones* 197
(i) *Francis North's theory*
(ii) *Roger North's experiment refuting that theory*

PREFATORY NOTE AND ACKNOWLEDGMENTS

The modern philosophy of music, as defined in Chapter 1 of this study, does not belong to a single discipline. For this reason, I sought to write in such a way as to be understood by a variety of readers. This task was made more difficult for a period—the seventeenth century—when new vocabularies, adequate to new philosophies, were in process of formation and when there was a dawning recognition of the need for unambiguous definitions of terms. In the 1650s the philosopher, Thomas Hobbes, provided a cogent argument for this need, when he asserted that definitions must be logically consistent with other definitions, as well as stated in the most efficient form possible.[1] But arguing for such definitions and achieving them are two different matters, as Hobbes well knew, for the making of concepts is a slow, tedious process that also relies on the cumulative efforts of numerous investigators.

As readers will discover, terms for temporal processes are central to the modern philosophy of music. During the seventeenth century, a number of these terms had yet to be given unambiguous definitions, so that recent interpreters have tended to substitute terms of later coinage that have such definitions—for example, 'frequency' for timing and 'phase' for relative timing. But one particular term—equality of timing—was given an unambiguous definition early in the century. Deriving from experiments with physical models, this concept guided the development of the modern philosophy of music, which, on the interpretation offered here, began in England with the 1677 publication of *A Philosophical Essay of Musick*.

As part of the evidence for this interpretation, Appendix A includes an edition not only of the *Essay* but also of three contemporary comments on it—one, by an anonymous individual, the other two, by Isaac Newton and Roger North, respectively. That of the last author appears in print for the first time, by courtesy of the British Library. My transcription, which in 1986 was the subject of a bibliographical project, was checked against the original manuscript by Mary Chan, who also assisted in a number of other ways, including comparing the physical features of three copies of the 1677 *Essay* now at Cambridge University.

The seed for this study was laid in 1988, when Alan Shapiro showed me a copy of Newton's comment, then in the Royal Society. Subsequently, this facilitated a comparison of the original text with three published transcriptions.

---

[1] See Hobbes, 'Elements of Philosophy'; see also Enriques, *The Historic Development of Logic*, pp. 70–76.

I also benefited from the generosity of a number of other busy people, with special thanks to John Bigelow for sharing interests in philosophical problems concerning time; David Tunley for close readings of a draft of this study; James Franklin for comments on some parts of the draft; my husband, Michael Kassler, for myriad helps including producing computer-generated illustrations and being on call when computer mishaps occurred; the publisher, Rachel Lynch, for patience and encouragement during the final stages of writing; and the desk editor, Sarah Charters, for finding gremlins that evade dimming eyesight.

Northbridge, New South Wales
Australia
October 2003

ABBREVIATIONS

In the editions that form Appendix A of this study, the original pagination or foliation of each text is indicated in bold type at the appropriate places in a text. In the study itself and in the footnotes to the editions, the original pagination/foliation, not the pagination of this book, is used for all references to these texts.

When manuscripts are cited, the symbols used refer to the library where the manuscripts are held. In abbreviations used for locations, the country is given first and then the library, as follows:

Australia (A)
- Nno Kassler Library, Northbridge, New South Wales

Great Britain (GB)
- Cjc St. John's College, Cambridge
- Cu University Library, Cambridge
- Lbl British Library, London
- Lpro Public Record Office, London
- Lrs Royal Society, London
- Ob Bodleian Library, Oxford

United States (US)
- Dlc Library of Congress, Washington, District of Columbia

Other abbreviations used in footnotes refer to (1) frequently mentioned members of the North family, (2) volumes in the *North Papers* series and (3) biographical writings of Roger North. Abbreviations used for the North family are given below, whereas those for the *North Papers* and for Roger North's biographical writings are given in the list of references at the end of this book.

NORTH FAMILY

DN (1) = Dudley North (1602–77), fourth baron North (1666)
- Father of nine children, author and amateur musician, he married on 24 April 1632 Anne, a daughter by the second wife of Charles Montagu [Montague], one of eight sons of Edward Montagu of Boughton. In the

early years of their marriage, the couple lived at Kirtling, Cambridgeshire, the seat of the third baron North. In 1638 they removed to Tostock in Suffolk, continuing to visit Kirtling. In 1646 DN (1) was named to a commission to investigate matters relating to the Norfolk Fens; and on May 1649 he was appointed one of the commissioners to assist with judging questions arising from the execution of the Act for draining the Fens. It is probable that this work brought him into contact with Jonas Moore (see Appendix B). He was buried on 27 June 1677; his will was proved on 30 October 1677 [(GB:Lpro) Prob 11/354]. For a modern biography, see Randall, *Gentle Flame*; for the Montagu connection, see Peter Millard in RN/*Life JN*, appendix iv.

FN = Francis North (1637–85)

- Second son of DN (1); for biographical details, see Chapter 1 *infra*. On 5 March 1672 FN married Frances, daughter and co-heiress of Thomas Pope, 3d earl of Downe. When her mother died in 1678, she and her two sisters inherited the family estate, Wroxton Abbey, near Banbury, Oxford. FN then bought the shares of the two sisters, so that the estate passed to him and his descendants. After his marriage, FN took lodgings in Portugal Row, Lincoln's Inn Fields, and then moved to a large house in Chancery Lane next to Serjeant's Inn. When his wife became ill, he took a country house at Hammersmith, where two of his brothers, JN and RN (see below), spent the nights, returning to London by day. She died on 15 November 1678, leaving three children: Francis, Anne and Charles. FN's portrait was twice painted by Peter Lely (see Appendix B). Two other portraits were painted in 1683, one by John Riley, the other by David Loggan, both of which are in the National Portrait Gallery, London.

DN (2) = Dudley North (1641–91)

- Third son of DN (1), he was educated at King Edward VI free grammar school in Bury St. Edmunds but dismissed about 1653, when he was sent to a writing school in London to study penmanship and accounting. Afterwards, he was apprenticed to a Levant merchant in London (1657–60), spending part of his apprenticeship in Smyrna (1661). He visited London briefly, before returning to Smyrna as a full-fledged member of the Levant Company (1662). Subsequently, he became a partner in a Constantinople trading house, and a few years later opened his own firm. By the age of thirty he was the most substantial of the English merchants and had acquired a command of Turkish and the dialects in use in the Levant. He returned permanently to England about 1680, residing briefly with FN in Chancery Lane, where the close friendship of their youth was revived. Subsequently, he formed a new friendship with his youngest brother, RN (see below), with whom he shared an interest in making experiments; see RN, *Life/DN*. On 12 April 1683 he married; two of three children survived to maturity. In 1691, after falling ill, he resided with RN, who assisted in bringing to completion, and writing a preface to, his *Discourses upon Trade*. He died at RN's house on 30 December 1691; his will was proved on 13 February 1692 [(GB:Lpro) Prob 11/409]. His portrait was painted by Peter Lely, as well as by Lely's pupil, Henry Tilson. For a modern biography, see Grassby, *The English Gentleman in Trade*.

JN = John North (1645–83)

- Fourth son of DN (1), he was educated at King Edward VI free grammar school in Bury St. Edmunds. Admitted to Jesus College, Cambridge, as a fellow commoner (1661), he was made a fellow by Royal mandate (1664). In 1672 he became professor of Greek in the University, thence migrating to Trinity College. In 1676 he was made chaplain to the king, and in 1677 master of Trinity. In 1680 he suffered a stroke, lingering on in a disordered state until his death in April 1683. His will was proved on 27 April 1683 [(GB:Lpro) Prob 11/372; see also 6 July 1683, Prob 11/374]. As a schoolboy in Bury St. Edmunds, his portrait was drawn by a friend of Peter Lely; see RN, *Life/JN*, p. 100; see further, Chapter 1 *infra*.

RN = Roger North (1651–1734)

- Youngest son of DN (1), he lived some years at his grandfather's house at Kirtling. Falling ill, he was educated privately and then at the free grammar schools in Bury St. Edmunds and Thetford. On 30 October 1667 he entered Jesus College, Cambridge, where he was tutored by JN. A second illness terminated his university studies. On 21 April 1669 he was admitted to the Middle Temple, London; on 29 May 1674 he was called to the bar and on 27 October 1682 to the bench. In the following year he became treasurer. His public appointments included steward to the See of Canterbury (1680), king's-counsel (1682), solicitor-general to James II (1685) and solicitor-general to the queen (1685). After the Revolution of 1688, when he was forced to resign his public offices, he removed to a house in the Piazza, Covent Garden, leased from the estate of Peter Lely (see Appendix B). In 1690 he purchased an estate in Rougham, Norfolk; and in 1692, after the funeral of DN (2), he began the work of remodelling. On 26 May 1696 he married Mary, daughter of Robert Gayer. In 1701 he made Rougham his permanent residence, renting out his rooms at the Middle Temple. He died 1 March 1734; his will was proved on 19 October 1734 [(GB:Lpro) Prob 11/667]. Of his seven surviving children, born between 1697 and 1712, the second son, Montagu, became custodian of his manuscripts and his first editor. A portrait of RN, painted by Peter Lely in 1680, is in the possession of his descendants. For his autobiography to c.1698, see RN, *Notes of Me*; for his contributions to the philosophy of music, see Kassler, 'Roger North'.

1 INTRODUCTION TO *A PHILOSOPHICAL ESSAY OF MUSICK* (1677)

In common parlance, the term 'philosophy of music' designates aesthetics, a branch of philosophy that emerged in the second half of the eighteenth century and that (depending on one's standpoint) concerns itself with musical beauty, musical taste or musical pleasure. In this study, however, the philosophy of music is understood more broadly as providing explanations concerning three kinds of temporal processes: the production, transmission and reception of acoustic signals, the action of the attention in listening to music, and the unfolding of music itself as it moves through time. As a counterpart to tonality—a new musical practice that emerged around 1600,[1] the philosophy of music, thus understood, was given its most definitive determination by Hermann von Helmholtz in the fourth and final edition of *Die Lehre von den Tonempfindungen als physiologische Grundlage für die Theorie der Musik*, published in 1877.[2]

In that work Helmholtz presented a detailed explanation of the phenomena of hearing that 'rested on a careful distinction between the events of the external world, the sensations (*Empfindungen*) they stimulate in our organs of sense, and the perceptions (*Wahrnehmungen*) of external objects or events which our mind makes on the basis of sensations'.[3] His explanation relied on a continuum theory in which sound waves are transmitted through an elastic medium; a resonance theory of hearing in which the ear performs a frequency analysis; an empirical theory of perception that relies on a principle of conscious decay under habituation;[4] and an exposition of the harmonic

---

[1] Tonality reached its culmination in art music at the end of the nineteenth century, although the musical 'language' continues in evidence today in some art music and, most notably, in commercial music. Wienpahl, 'Modality, Monality and Tonality', outlines the emergence of tonality in a statistical study of 5, 179 items of music composed between 1500 and 1700. The OED gives 1838 for the first English use of the term; for its earlier introduction in Europe and for the wider context, see Hyer, 'Tonality'.

[2] For an account of some of the modifications made by Helmholtz to the different editions (1863, 2/1865, 3/1870, 4/1877), see Boring, *Sensation and Perception*, pp. 404–406, 432; for other modifications made by the English translator, Alexander Ellis, see various of his notes to Helmholtz, *On the Sensations of Tone*.

[3] Maley, *The Theory of Beats*, p. 129; see also pp. 135–136.

[4] For Helmholtz's principle, see pp. 125–126 *infra*.

series and its importance for the science of music.[5] Although Helmholtz's chief object was to explain the physical and physiological basis for music, he also explored the implications of his findings for aesthetics and for the development of the feeling for tonality in European musical history.

Prior to 1677 no English writer published or translated into English a treatise devoted solely to the philosophy of music, taken in this broad sense, even though the subject had received extensive treatment on the continent, most importantly by the French priest, Marin Mersenne. But Mersenne's work was Janus-faced: on the one hand, he was forward looking in his conviction that a rational explanation of music was possible when firmly based on sense experience and experiment; on the other hand, his actual writings were humanistic and encyclopaedic, thereby reflecting Renaissance ideals.[6] England, however, seems to have been isolated from the encyclopaedic tradition.[7] Instead, parts of the philosophy of music were included either in books on natural philosophy or in tracts on musical practice.[8] Other parts were explored in experimental investigations, some of which were never published.[9] And still other parts had to wait some considerable time before being treated at all.

There are a number of reasons for this piece-meal approach to the philosophy of music. First, acoustics had yet to emerge as an autonomous science, as appears from one of the Lucasian lectures on mathematics, read by Isaac Barrow in 1664, where there is the conditional judgment that

---

[5] The term 'harmonic series' denotes a set of frequencies, each component of which is an integral multiple of the fundamental frequency, whereas the term 'frequency' denotes the rate of recurrence of a regularly repeated event (e.g., a vibration), or the number of times that it occurs in an assumed unit of time.

[6] Although Mersenne was a transitional figure, as Dear, 'Marin Mersenne', recognised, his modernity has been stressed by Lenoble, *Mersenne*; Crombie, 'Marin Mersenne'; Cohen, *Quantifying Music*, pp. 97–114; and others. The introduction of Mersenne's work into England has yet to be studied, although it seems that John Wilkins did much to popularise some aspects of it; see Aarsleff, 'John Wilkins'. Robert Hooke, who was closely associated with Wilkins, owned a number of Mersenne's writings, including two works with important contributions by Thomas Hobbes; see Rostenberg, *The Library of Robert Hooke*. For Mersenne's publications relating to Hobbes, see Shapiro, 'Kinematic Optics', p. 145, and Kassler, *Music, Science, Philosophy*, p. 86.

[7] See, e.g., Kassler, *Music, Science, Philosophy*, pp. 165–200.

[8] The tracts on musical practice have been the subject of a number of monographs and studies, some of which are cited in this book; but the books on natural philosophy have so far eluded detailed investigations along the lines of Cohen, *Quantifying Music*.

[9] This statement applies in particular to Hooke, a central figure in experimental investigations relating to the philosophy of music. For an assessment of some of his work, see Gouk, 'The Role of Acoustics and Music Theory'; Kassler and Oldroyd, 'Robert Hooke's Trinity College "Musick Scripts"'; Kassler, *Inner Music*, pp. 124–159, and *Music, Science, Philosophy*, pp. 111–124.

> if the Figure of the Air's Undulation could be the same way discovered, by which Sound is performed, and the Sense of Hearing impelled, there would doubtless arise thence a new Part of Mathematics to be celebrated by the Name of *Acustics*, or the Science of Sounds.[10]

Second, a theory of auditory resonance would have to await developments leading to the understanding of the structure and function of the ear, as well as of the mechanics of vibration. Third, although in the 1650s Thomas Hobbes developed an empirical theory of perception that relied on a principle of conscious decay under habituation, this theory would not be widely accepted, in part because of the heterodox nature of Hobbes's writings.[11] Fourth, although harmonics had been recognised as a problem, because they implied that different modes of vibration occur simultaneously in a sound source, an understanding of this problem would not begin until harmonics came to be associated with the existence of nodal points on a vibrating string.[12]

Finally, there seemed to many that an unbridgeable 'gap' existed between that part of the philosophy of music treating acoustic signals in the external world and that part treating the human receiver and knower.[13] This gap is best exemplified by a brief examination of the way in which Galileo Galilei, early in the seventeenth century, dealt with 'musical' problems in his *Discorsi e dimostrazioni mathematiche, intorno à due nuoue scienze* (1638).[14] There, he applied the constancy of the period of pendulum oscillations to the phenomena of hearing, since he also understood these phenomena as having to do with rates of vibration. The intervals of vibration observed when the action of the pendulum is investigated provide a clue for the solution of the ratios of musical intervals, for Galileo supposed that vibrations from a sound source strike the eardrum, causing it to vibrate in the same time as the source and to produce the stimulus that 'the mind translates into sound'.[15] The effect of a musical interval may be smooth or rough, pleasant or unpleasant, so that it is an affect or inward disposition of the listener. But the question of affection is demarcated from the pendular vibrations providing the stimulus to the ear, since Galileo gave no indication that there is any further problem to be solved in respect to the mechanics of the vibrations. What concerned him was the connection between the vibration rates and the ratio of the heard intervals.

---

[10] Barrow, *The Usefulness of Mathematical Learning*, p. 24.

[11] For Hobbes's principle, see Kassler, *Inner Music*, pp. 91–93, 108–112.

[12] See Dostrovsky, 'Early Vibration Theory', pp. 193–196, 204, 207, 209.

[13] See Cohen, *Quantifying Music*, pp. 2, 31–32, 36, 81, 110–111 *et passim.*

[14] Galilei, *Two New Sciences*, pp. 42–43, 94–108.

[15] Galilei, *Two New Sciences*, p. 101.

The explanation of why certain intervals make a pleasant impression involves more than the mechanics of the vibrations producing the stimulus; and if followed up, it leads to questions that are not mechanical questions. For a philosophy of *music*, however, such a follow-up is required. This had been the aim of Helmholtz, who supposed that previous to his work the boundaries of acoustics and musical science had remained distinct, that acoustics, while of service to musical instrument makers, had been barren for the development of the theory and aesthetics relating to tonality.[16] Helmholtz, therefore, made the physiology of hearing the connecting link between physical acoustics and musical science. Not long afterwards, and certainly by the time of his death in 1894, musical practice was moving 'ever further from the harmonic canons' that Helmholtz's theory of hearing explained.[17] At the same time, new models of the physiology of hearing began to undermine Helmholtz's physical model.[18] Nevertheless, these and other changes do not indicate that Helmholtz failed to describe and explain the musical practice called 'tonality'. Rather, they suggest the importance of understanding his achievement historically, as the climax of a particular tradition—the modern tradition in the philosophy of music.[19]

This study is intended as a step towards that understanding, since the philosophic tradition that reached its culmination in the work of Helmholtz begins in a modest way in 1677 with the publication of *A Philosophical Essay of Musick*, in which an anonymous author attempts to link physical acoustics and musical science, thereby providing the first rationale for tonality. In the following chapters I examine that *Essay* within its historical context, consider three contemporary criticisms of it and then outline how one of the critics developed some of the implications. As readers will discover, the anonymous author argues his case from a conjecture—an hypothesis concerning the generation ('cause') of sound. In this chapter, therefore, I provide a brief outline of the state of hypothetical physics between 1650 and 1676, insofar as that physics has to do with the phenomena of hearing—sound and music. I then introduce readers to the author, with particular attention to the genesis of the *Essay* and the friend to whom it is directed.

---

[16] Helmholtz, *On the Sensations of Tone*, p. 1.

[17] Maley, *The Theory of Beats*, pp. 135–136.

[18] For some of these models, see Tonndorf (ed.), *Physiological Acoustics*, pp. 98–156; for some implications, see, e.g., Winckel, *Music, Sound and Sensation*.

[19] The term 'modern', as used in this study, refers to the period from the seventeenth century to the closing decades of the nineteenth century, whereas the term 'modern philosophy of music' refers to a narrower period, 1677 to 1877.

HYPOTHETICAL PHYSICS 1650–76
I. 1650 TO 1660

During the seventeenth century the propagation of sound was explained by recourse to one or the other of two hypotheses—an emission or a continuum hypothesis.[20] The emission hypothesis, deriving from the ancient atomists, was developed by the modern Epicureans, who believed that in the propagation of sound there is a transport of matter from a sounding source to the ear. This hypothesis had at least two variants. In one, sound was attributed to the emission of minute, invisible particles,[21] whereas in the other, sound was attributed to the emission of invisible 'textures'. The second variant was indebted to a Frenchman, Pierre Gassendi, whose philosophy was based on the premise that the world consists of very small spaces not filled by any body and very small bodies that have within them no empty space.[22] In 1654 Walter Charleton produced a compendium of Gassendi's philosophy for English readers, in which sound, '*a Corporeal Ens*', is defined as

> the subtler or more æthereal part of aer, extrite and formed into many small *(Moleculæ)* masses, or innumerable minute Contextures, exactly consimilar in Figure, and capable of affecting the Organ of Hearing in one and the same manner: which configurated small masses of aer fly off from bodies compulsed or knockt each against other, with some violence; and progress by Diffusion in round.[23]

On this definition, sound is a material structure consisting of a heap of minute particles ('contextures') that are similarly shaped ('configurated'), so that when a sonorous source is percussed, the emitted textures are transported through the air to the ear as a stream ('diffusion'). Charleton supposed that, within the sphere of a texture's stream, '*excentral* lines', emitted from a sounding source at the centre of the sphere, form a cone with their base on that

---

20 I follow the terminology used by Shapiro, 'Kinematic Optics', p. 136 n.5, who argued that the terms 'corpuscular' (for emission) and 'wave' (for continuum) mislead when applied to the seventeenth century.

21 It is possible that Isaac Beeckman developed this early variant; see Cohen, *Quantifying Music*, pp. 120–123.

22 See Rochot, 'Pierre Gassendi'; see also, Gassendi, 'Manu-ductio ad theoriam, seu partem speculativam musicæ', *Opera omnia*, vol. 5, pp. 631–658. Cohen, *Quantifying Music*, p. 285, suggests that the treatise was written between 1650 and 1655. For the 'new' Epicureanism and its transmission to England, see Jones, *The Epicurean Tradition*, pp. 166–213.

23 Charleton, *Physiologia*, pp. 212–213. To date, there is no systematic comparison between Gassendi's writings on the phenomena of hearing with the relevant portions in Charleton's compendium.

body and their vertex in the listener's ear [*Figure 1.1*].[24] He further supposed

Charleton's 'stream lines' from (a) the sounding source to (b) the ear

Figure 1.1

that each sound requires its own specific texture so that, in sense perception, the differences between one sound and another may be distinguished; for, when different textures reach the ear, they enter its duct and strike the auditory nerve. Those textures that accommodate the pores of the nerve make a gentle, smooth or equal impression on the nerve's internal filaments, whereas those that do not accommodate the pores make a rough, harsh or unequal impression.[25] In music, for example, the most grateful texture is the unison, because 'as the Vibrations of the strings, so will the strokes inflicted on the sensory, have the same proportion each to other, as one hath to one (the proportion of Equality) and consequently will be equal in number and time, so as to affect the sensory most equally and Unitedly'.[26]

---

[24] Charleton, *Physiologia*, p. 210.

[25] Charleton, *Physiologia*, p. 216.

[26] Charleton, *Physiologia*, p. 225.

This type of explanation is now referred to as the coincidence theory of harmony,[27] according to which concord is related to rates of vibrations—frequencies—of the source of sound. These rates could be increased or decreased by tensing or relaxing a musical string, as Charleton took pains to describe. But, on his variant of the emission hypothesis, ideas of consonance have a material cause: bodies when percussed emit textures—configurated masses in motion—that stream into the ear and strike the auditory nerve. When the percussions on the nerve coincide frequently in agreement, the result is concord, and when the percussions coincide infrequently or break in on one another, the result is discord. According to this supposition, strings vibrating in the ratio 2:1 would coincide on every other vibration; in the ratio 3:2, on every sixth vibration; and so on. In short,

> the Consonance is by so much more grateful, by how much more frequently the strokes unite in the Sensory: and *è contra*. Whence is generated the Dissonancy, or ingratefulness of Sounds, when ever the strokes either too rarely, or never unite: because, in those cases, the sense is held in a kind of lasting distraction, and unless a restitution of the distracted parts of the Sensory be made by some Coalitions, and those sufficiently frequent (which are a kind of Balsam, to cure the gratings and dissolutions) the sensory must be mis-affected with a kind of Laceration, and undergo that dolour unwittingly.[28]

Charleton's coincidence theory, therefore, is to be understood within the context of his emission hypothesis, which posits a transport of matter.

In the competing hypothesis—the continuum hypothesis—there is no transport of matter; rather, what is propagated is a state through an intervening medium. This hypothesis, traceable to Aristotle, was developed by the scholastics, who explained the propagation of sound by the concept, multiplication of species, where the term 'species' denoted immaterial but moving patterns, like vortices (eddies), that represent the continuous action of the transmuted 'power' of percussion of a dense body on a rarer one. Such action, though conceived as incorporeal, invisible and 'spiritual', by moving corporeal 'instruments' at any time becomes corporeal, visible and 'material'.[29] Thus, when a bell (dense body) is struck, the force of percussion is immediately 'sensed' in the whole body (the bell), which, in turn, strikes the air

---

[27] I do not know when the name was first coined. For the most detailed examination of the theory to date, see Cohen, *Quantifying Music*, who ignores Charleton; moreover, his statement, p. 177, that Gassendi gave a 'wave' account of the propagation of sound is incorrect (no such account is possible on an emission theory).

[28] Charleton, *Physiologia*, p. 225.

[29] See, e.g., Kenneth Keele, *Leonardo da Vinci*, p. 215.

(rarer body), causing it to 'resound' as a whole—that is, to move to and fro like a fan, thereby striking the eardrum (another dense body), which then resounds. Dense bodies react like that of a percussed bell, so that a resounding dense body can make another dense body that is similar respond in sympathy.

Philosophers who rejected the scholastic explanation but who still wished to retain a continuum hypothesis could choose between two main variants, static and kinematic, that emerged during the first half of the seventeenth century. In his first treatise, the *Compendium musicæ*, René Descartes developed the static variant along Pythagorean lines.[30] Although preferring to focus not on physics but on the ease with which the listener distinguishes certain arithmetical proportions,[31] Descartes hinted that when the organ of hearing is struck by an outside stimulus that is regular, sense perception occurs if the organ takes on the ratio of the perceived object. Subsequently, Descartes began to fill in the details of his philosophy, in which he identified space with matter and argued that the propagation of all action is by contact between particles of this matter.[32]

Then, in *De homine* he provided a physical account of the phenomena of hearing, according to which sound is defined not as a motion (which takes time) but as an inclination or tendency to motion (*conatus*), whereby an impact, or series of impacts, is transmitted rectilinearly by an internal push of micro-bodies—invisible air particles that fill the interstices of macro-bodies that are visible.[33] For the 'idea of sound', the disturbance (push) must reach the eardrum, the air particles on the other side of that membrane, the even more rarefied, subtle particles ('animal spirits') within the pores of the auditory nerve as well as those within the brain. There, the push 'impresses'—that is, spatially reconfigures—the imagination, like a seal on wax, so that the mind judges the degrees of smoothness or harshness from the more or less equal impressions on the imagination.

Like Charleton, Descartes gave primacy to the unison as the most pleasing musical sound, explaining this and other intervals physically in terms of 'little blows' as follows:

---

30 Written in 1618, Descartes' *Compendium musicæ* was published posthumously in Latin in 1650; the first English translation was issued in 1653.

31 Descartes, *Excellent Compendium*, p. 23.

32 See Gaukroger, *Descartes*, pp. 225–292.

33 Descartes, *Treatise of Man*, pp. 45–48, 71–72. This description is a consequence of Descartes' reliance on hydrostatics, a branch of physics that treats the pressure and equilibrium of fluids at rest. But Descartes denied the proper concept of pressure, which demands that at any point within a fluid the pressure acts equally in all directions; see Shapiro, *Light, Pressure, and Rectilinear Propagation*, p. 240.

> when many [blows] succeed one another, as one sees in the vibrations of strings and bells when they ring, then these little blows will compose one sound which [*a*] the soul will judge [to be] smoother or rougher according as the blows are more or less equal to one another, and [*b*] it will judge [to be] higher or lower according as they succeed one another more promptly or tardily, so that if they are a half or a third or a fourth or a fifth more prompt in following one another, they will compose a sound which the soul will judge to be higher by an octave, a fifth, a fourth, or perhaps a major third, and so on. And finally, several sounds mixed together will be harmonious or discordant according as more or less orderly relations exist [among them] and according as more or less equal intervals occur between the little blows that compose them.[34]

Here, then, is Descartes' version of the coincidence theory of harmony.[35] But a caution is necessary, since his version is to be understood as considering the total number of coincidences between 'blows' *without regard to time*.[36] This is because Descartes' approach was causal, not temporal, so that the changes described externally to, and internally within, the human body are instantaneous.

In the *Compendium musicæ* as well as in *De homine*, Descartes aimed to establish absolute guarantees for the veracity of sensory perception and, hence, for objectivity. For him, there were two guarantees, both measurable. One is found in the perfect arithmetical proportions—the Pythagorean simple numbers 2, 3 and 5. The other is found in a mechanised theory of sensory perception, which Descartes conceived as instantaneous physiological changes (reflexes) within the auditory (ear-brain) system. For Descartes, however, 'body'—including the auditory system—is autonomous and self-regulating, so that it functions independently of 'mind'. Hence, he described the rapport between body and mind solely in terms of arithmetical proportions. But in transforming those proportions into visible line segments of different lengths, he geometrised the arithmetical proportions. As a consequence, Descartes was led to equate the phenomena of hearing with what in effect was his geometrical model. This equation implies two things. First, in sensory perception, the spatially reconfigured imagination represents line lengths, not arithmetical numbers or ratios. Second, in perceptual cognition, mind, from the light of

---

[34] Descartes, *Treatise of Man*, p. 47.

[35] Descartes, *Excellent Compendium*, p. 23, hinted that interval ratios depend on the coincidence of 'percussions or verberations'; but this hint appeared as an explanation why major thirds are the most perfect intervals and only after Descartes had offered a geometrical proof, using a line-length diagram; see Green, *The Harmonic Series*, pp. 147–148.

[36] Cohen, *Quantifying Music*, pp. 174–175, did not treat the problem of time in Descartes' coincidence theory.

reason, grasps the appropriate rational proportions from the represented line lengths.[37]

Galileo took a different approach, when he conceived the propagated state as an actual motion—pulse, vibration, wave—that takes place in time. In England, Hobbes developed this second, kinematic variant of the continuum hypothesis, for he was a fierce critic of the scholastics as well as of Descartes. To Hobbes, the scholastic 'opinion' that sound and noise are the qualities of the bell or the air

> hath been so long received, that the contrary must needs appear a great paradox; and yet the introduction of species ... (which is necessary for the maintenance of that opinion) passing to and fro from the object, is worse than any paradox, as being a plain impossibility'.[38]

In 1650, when Hobbes made this statement, the old, scholastic philosophy was losing favour. But this was not the case with Descartes' philosophy, which was still relatively new in England. To undermine that philosophy, Hobbes described a propagated disturbance not as a push but as a pulse—a vibration or 'reciprocal motion'—through an all-pervading medium. His work, therefore, is important as an early attempt to understand the motion involved in wave propagation.[39]

But Hobbes's philosophy has an additional importance, since it introduced to English readers Galileo's two new sciences: the sciences of resistance and pure motion—in modern terms, elasticity and kinematics, the latter of which is an extension of geometry by the introduction of the idea of time and the consequent idea of velocity. Moreover, the influence of Galileo's thought may be gleaned from three mechanical principles that constitute the basis of Hobbes's philosophy.[40] First, endeavour (*conatus*) is not an inclination or tendency to motion, as Descartes supposed; rather, it is an actual motion in an indefinitely small (i.e., infinitesimal) interval of time—'motion through space and time less than what is given, that is to say, less than can be determined by exposition or by number; that is[,] motion made through length of a point, and in an instant or point of time'.[41] This is the main concept of

---

[37] For details, see Gaukroger, *Descartes*.

[38] Hobbes, 'Human Nature', p. 4. Note that Hobbes's criticism of the scholastics would apply equally to the emission hypothesis of Charleton and other modern Epicureans.

[39] See Hobbes, 'Elements of Philosophy', pp. 485–501.

[40] See Kassler, *Inner Music*, pp. 49–107, and *Music, Science, Philosophy*, pp. 82–124; see also Shapiro, 'Kinematic Optics', who was the first to draw attention to Hobbes as the originator of kinematic optics modelled on acoustics.

[41] Hobbes, 'Elements of Philosophy', p. 206.

Hobbes's philosophy.[42] Second, the idea of resistance is closely related to the endeavour concept, since Hobbes's model of impact is that of an elastic process entailing deformation and restitution through continuous intervals to and through rest.[43] For him, equilibrium (rest) is thoroughly kinetic, because it conceals an internal or dynamic tension of competing forces. Third, physical systems are conserved by the dynamic inner tension that equalises action and reaction.

These principles are set out in the first of three volumes that constitute Hobbes's last complete statement of his philosophy, two of which—*De corpore* and *Human Nature*[44]—include treatments of sound and music. In *De corpore*, the chapter on the physiology of the senses is devoted chiefly to the phenomena of hearing,[45] where Hobbes defined sound as

> *sense generated by the action of the medium, when its motion reacheth the ear and the rest of the organs of sense* [i.e., the ear-brain system]. Now, the motion of the medium is not the sound itself, but the cause of it. For the phantasm which is made in us, that is to say, the reaction of the organ, is properly that which we call *sound*.[46]

On this definition the auditory system is a reaction system. How, then, is sound generated and propagated to that system? According to Hobbes,

> in the generation of sound, the first collision or breaking asunder beateth away and driveth out of its place the nearest part of the air, and that to a considerable distance, and with considerable velocity; and as the circles grow by their remoteness wider and wider, so the air being more and more dissipated, hath also its motion more and more weakened.[47]

---

42 See Kassler, *Inner Music*, pp. 82–85.

43 Hobbes, 'Elements of Philosophy', p. 211: 'I define resistance to be the endeavour of one moved body ... contrary to the endeavour of another moved body'; OED 3.b. cites this definition as the first English use of the term 'resistance' in the physical sciences.

44 Because of his illness and other factors, the publication history of these volumes differs from Hobbes's planned systematic order for his philosophy. This order, for the two volumes in question, should have been: (1) 'Elements of Philosophy', i.e., *De corpore* (Latin 1655, English 1656), on first philosophy—body or body natural—leading to (2) *Human Nature* (1650). See Macdonald and Hargreaves, *Thomas Hobbes*; for Hobbes's illness, see Kassler, *Music, Science, Philosophy*, pp. 83–85.

45 Hobbes, 'Elements of Philosophy', pp. 485–508.

46 Hobbes, 'Elements of Philosophy', p. 485.

47 Hobbes, 'Elements of Philosophy', p. 497.

In short, sound is produced by the displacement of air particles, and it is propagated like surface waves on water.[48] Air is the medium of sound—but not the only medium, for Hobbes asserted that 'water, or any other body how hard soever, may be that medium'.[49]

What happens, then, to the reaction system that gives us a sense perception ('phantasm') of sound? To explain this, Hobbes relied on a geometrical representation to demonstrate how the disturbance—propagated 'in equal and equally frequent spaces of time' from two strings vibrating in the same time— strikes the 'sides' of the eardrum, thereby producing a pleasing sensation ('concent') of the unison ('the greatest concord'), octave, fifth and so on. In this demonstration [*Figure 1.2*], A B and C D represent two equal strings that begin vibrating at the same instant; E F and G H represent the 'latitudes of their vibrations';[50] and the points E G F and H represent the 'concave superficies' of the eardrum.

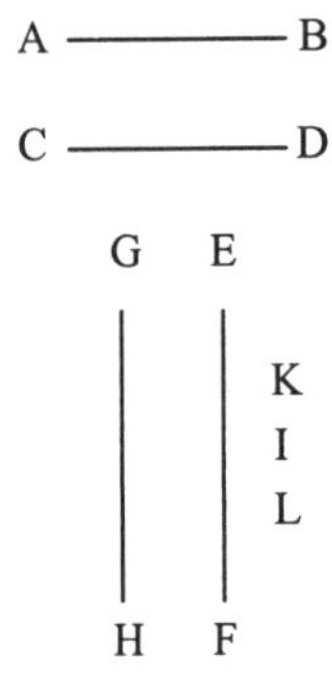

Hobbes's demonstration of sound reception

Figure 1.2

---

[48] Hobbes, 'Elements of Philosophy', p. 487: 'the first motion diffuseth itself every way into circles, such as are made by the throwing of a stone into a standing water, where the velocity grows less and less as the undulation proceeds further and further from the beginning of its motion'.

[49] Hobbes, 'Elements of Philosophy', p. 498.

[50] For the term 'latitude', see OED *sb.* 2.a. *rare* extent, range (1605–1851). But the phenomena of hearing are temporal phenomena; hence, Hobbes's 'latitudes' denote spaces of time through which the motion of the air is propagated during one vibration cycle of the sounding source.

For the interval of the fifth, Hobbes's account is as follows:

> the vibration of the string A B remaining still the same it was, let C D be stretched till its vibration be swifter than that of the string A B in the proportion of 3 to 2, and let E F be divided into three equal parts in K and L. In what time therefore the string C D makes one third part of its vibration, which third part is from G to H, the string A B will make one third part of its vibration, that is to say, two-thirds of E F, namely, E L. And in what time the string C D makes another third part of its vibration, namely H G, the string A B will make another third part of its vibration, namely from L to F, and back again from F to L. Lastly, whilst the string C D makes the last third part of its vibration, that is from G to H, the string A B will make the last third part of its vibration from L to E. But the points E and H are both in the sides of the organ [i.e., the eardrum]. Wherefore, at every third time the organ will be stricken by the vibration of both strings together, and make the concord which is called a *fifth*.[51]

This, then, is Hobbes's version of the coincidence theory, which is similar to that of Galileo, who, in 1638, supposed that the correlation of pitch with rates of vibration was an established fact. Hence, he believed that he could provide an explanation why certain pairs of tones differing in pitch produce a pleasing sensation, others a less pleasing one and still others a disagreeable sensation, such an explanation being 'tantamount to an explanation of the more or less perfect consonances'.[52] His explanation was that the consonance of an interval is determined by the number of pulses of the higher tone that coincide with the number of pulses of the lower tone; and, unlike Descartes, he placed intervals in an order of consonance based on the time required for the consonance to achieve full cycle. Hence, he laid stress on pulse rates that are in time or out of time. For example, when 'vibrations of two different tones' strike the eardrum 'out of time [*sproporzionatamente*]', harsh sensations arise from the dissonance between tones whose ratios of vibration are incommensurable. But when vibrations from pairs of tones strike the eardrum 'with a certain regularity', smooth sensations arise from the consonance between tones whose ratios of vibration are commensurable. Galileo therefore conceived an interval like the fifth as a characteristic periodicity—three pulses against two—repeating at high speed; but because of its speed, the periodicity is 'insensible', that is, unavailable to conscious pattern recognition.[53]

---

[51] Hobbes, 'Elements of Philosophy', pp. 500–501.

[52] Galilei, *Two New Sciences*, p. 103.

[53] Galileo thus evades the problem that the sequence of pulses will not be synchronised to produce coincidences; see Dostrovsky, 'Early Vibration Theory', p. 182, and Walker, *Studies in Musical Science*, pp. 31, 32–33.

Hobbes specifically named Galileo in *Human Nature*, where he defined knowledge as evidence of truth 'from some beginning or principle of *sense*',[54] and where he established four principles of knowledge, the first of which is 'that we have such and such *conceptions*', that is, experience of facts.[55] But he was especially concerned

> to search out and declare *from what* conception *proceedeth* every one of those *passions* which we commonly take notice of: for, seeing the things that please and displease, are innumerable, and work innumerable ways, men have not taken notice but of a very few, which also are many of them without a name.[56]

To accomplish this aim, Hobbes provided a temporal framework for conception by distinguishing its three modalities: conception present, which is '*sense*'—that is, immediate sensation; conception past, which is '*remembrance*'; and conception future or '*expectation*'. Then, when considering the pleasures of hearing, he applied this framework to music, beginning with conception present.

For example, he argued that in the case of 'simple sounds'—tones taken individually—please by 'an equality [of timing] continued by the percussion of the object upon the ear' and displease when inequality of timing produces '*harshness*, such as is grating'.[57] But in the case of harmony, the many sounds 'together agreeing' please 'by the same reason as the *unison*, which is the sound of equal strings equally stretched', as well as by

> *inequality and equality alternate*, that is to say, the higher note striketh twice, for one stroke of the other, whereby they strike together every second time; as is well proved by Galileo, in the first dialogue [of *Two New Sciences*] concerning local motion: where he also sheweth, that two sounds differing a *fifth*, delight the ear by an *equality* of striking *after two inequalities*; for the higher note striketh the ear thrice, while the other strikes but twice. In like manner he sheweth wherein consisteth the pleasure of concord, and the displeasure of discord, in other difference of notes.[58]

---

[54] Hobbes, 'Human Nature', pp. 26–30, p. 28.

[55] For conception—obscure and distinct, see Hobbes, 'Human Nature', pp. 9–26.

[56] Hobbes, 'Human Nature', pp. 34–35.

[57] Hobbes, 'Human Nature', pp. 35–36, who also mentioned noises that result from non-periodic pulses, that is, 'sounds, which do not always affect the body, but only sometime, and that with a kind of horror beginning at the teeth.

[58] Hobbes, 'Human Nature', p. 36.

When sounds 'consisteth in *consequence of one note after another, diversified* both by *accent* and *measure*', 'that which pleaseth is called an *air*'. In this case, the pleasures of hearing derive not only from conception present but also from conception past (remembrance), so that there is a comparative act of the intellect. But in considering 'for what reason one succession in tone and measure is more pleasing than another', Hobbes confessed: 'I know not; but I conjecture the reason to be, for that some of them imitate and revive some passion which otherwise we take no notice of, and the other not, for *no air pleaseth but for a time*, no more doth imitation.'[59] Finally, Hobbes noticed a pleasure that 'happeneth only to men of skill in music', which is the 'rejoicing of their own skill'.[60] This pleasure, he wrote, derives from conception future (expectation), 'that is to say, in conception of power past, and the act to come'.

II. 1660 TO 1676

The philosophies of Charleton, Descartes and Hobbes represent the state of the new hypothetical physics in England to about 1660, insofar as that physics relates to the phenomena of hearing. Hence, by the 1660s English proponents of Hobbes's kinematic variant of the continuum hypothesis had to compete with only two alternative hypotheses: the continuum hypothesis in its static variant presented by Descartes, and the emission hypothesis in the Gassendian variant presented by Charleton. It is important to note, therefore, that despite the different approaches of Descartes and Charleton, whereby one argued for a plenum and the other for a void, their conceptual schemes of matter in motion differed very little. On the one hand, both men regarded material substance as absolutely homogeneous, so that they both accounted for the apparent variety in the world by the mere differences of shape, size, position and arrangement of invisible particles. They also treated fluids like air as composed of invisible particles and pores that either are filled with some subtle matter (Descartes) or are empty of matter (Charleton). On the other hand, both men restricted motion to the mechanics of simple machines and did not include kinematics.[61]

Although Robert Boyle seems to have preferred the emission hypothesis, his so-called 'corpuscular philosophy', developed from the 1660s, includes descriptions of the phenomena of hearing utilising either a Charletonian or a

[59] Hobbes, 'Human Nature', p. 36, who followed this statement with the supposition that colours might give pleasure in the same way as sounds.

[60] Hobbes, 'Human Nature', p. 37.

[61] See Dijksterhuis, *The Mechanization of the World Picture*, p. 417, and Westfall, *Force in Newton's Physics*, pp. 99–109, 138–139, 535–537.

Cartesian scheme.[62] Thus, if one body (fluid or solid) strikes another and we hear a sound of a particular pitch, this is because textures—configured masses and motions of invisible corpuscles that, for example, produce tensions in musical strings—are modified. These modifications are propagated either as effluvia (streams) or as disturbances (pushes) by means of corresponding modifications in the texture of the external air corpuscles surrounding bodies, the texture of internal air corpuscles surrounding the sensitive parts of the ear and the texture of the more subtle corpuscles within the nervous system and within the brain.[63]

Until 1686, when Isaac Newton introduced attractive and repulsive forces acting on particles, particle collision was the basic phenomenon of seventeenth-century physics. But there were different ways of treating the mutual action of bodies in collision, some philosophers concerning themselves solely with the forces involved in collisions and percussion, others, with the problem of the communication of motion from one body to another.[64] These different treatments depended on whether bodies were conceived as elemental, invisible and hypothetical, or as bodies of ordinary experience. They also depended on whether such physical properties as inelastic hardness or flexibility were assigned to one or both kinds of bodies. In the 1668 investigations promoted by the Royal Society, for example, Christopher Wren and Christian Huygens treated elemental bodies as perfectly elastic, so that in the collision of such bodies, no motion is lost and the sum of the two products of 'body' and velocity is constant.[65] John Wallis, however, equated elasticity with less-than-perfect hardness and hardness with lack of elasticity, so that in collision, such bodies will not rebound and motion will be lost, not conserved.[66]

By the 1650s Hobbes had observed the effects of elasticity in the motion of restitution; hence, he argued that, in the deformation of matter, motion is conserved, because macroscopic motion is translated into microscopic and manifested as heat.[67] And, in conceiving collision as an elastic process

[62] The term 'corpuscular' was given general currency through the influence of Boyle; see Alexander, *Ideas, Qualities and Corpuscles*.

[63] Boyle's works contain some hints about his own musical background. For example, in 1671 he wrote: 'I ... am very musically given', and in 1674 he referred to his music master, though without naming him; see Boyle, *The Works*, vol. 2, p. 181, vol. 3, p. 801; see also Kassler, *The Science of Music in Britain*, vol. 1, pp. 111–114, and Kassler, *Music, Science, Philosophy*, pp. 111–124.

[64] Kalmar, 'Thomas Hariot's *De Reflexione Corporum Rotundorum*, p. 201.

[65] Hall, 'Mechanics and the Royal Society, 1668–70', pp. 28–37.

[66] Scott, *The Conflict between Atomism and Conservation Theory*, pp. 3–21.

[67] For Hobbes's kinetic theory of heat, see Kassler, *Inner Music*, pp. 69–70.

equalising action and reaction, Hobbes relied on physical models such as stretched strings.[68] By contrast, philosophers who conceived bodies as inelastic treated collision on the Aristotelian analogy of the motion of a ball after impact.[69] According to Alan Shapiro, the ball analogy, which was prevalent throughout much of the seventeenth century, was used to account for both sound and light, including their deviation in reflection and refraction from a rectilinear path. The aerial propagation of sound presented a more difficult problem for, unlike light, sound could not be seen; and unlike the supposed rectilinear light rays, sound seemed to deviate in many different ways from a rectilinear path. Hence, some philosophers appealed to terms such as 'spherical lines' to describe the observed deviations, whereas others drew an analogy to water waves, which do not travel in straight lines.[70]

The water-wave analogy was very widespread; but the wave surfaces themselves received scant attention, in spite of Barrow's remark of 1664, quoted previously, that

> If the Figure of the Air's Undulation could be the same way discovered, by which Sound is performed, and the Sense of Hearing impelled, there would doubtless arise thence a new Part of Mathematics to be celebrated by the name of *Acustics*, or the Science of Sound.[71]

To discover 'the Figure of the Air's Undulation' required knowledge of the air itself, since on the emission hypothesis the material cause of sound is the most rarefied or subtle part of aerial matter and on the continuum hypothesis the efficient cause of sound is a disturbance in the aerial (or other) medium. What, then, of the nature of the air? In 1673 some answers to this question were included in a report by John Wallis, who found that, by the term 'air',

> Mr. *Hobbes* would sometimes (as in *Dialogus Physicus* p. 4) have us understand a *pure Æther* ... which, I suppose, answers the *Materia subtilis* of *Des Cartes*, and Mons. *Hugens* [sic] his more *subtile Matter* than *Air*. On the other hand, M. *Hugens* here, by *Air*, seems to understand that feculent matter arising from those the *Earths and Waters Effluvia*, which are intermingled with this *subtile Matter.*[72]

---

[68] See Kassler, *Inner Music*, pp. 49–107, and *Music, Science, Philosophy*, pp. 83–124.

[69] For Descartes' use of this analogy, see Eastwood, 'Descartes on Refraction'; for an approval of the analogy, see Charleton, *Physiologia*, p. 214.

[70] Shapiro, 'Kinematic Optics', pp. 136–137.

[71] Barrow, *The Usefulness of Mathematical Learning*, p. 24.

[72] Wallis, 'An Extract of Letters', p. 5161. For Hobbes's æther, see Kassler, *Inner Music*, pp. 64–66, 69–70, 80, 82, 85, 118, 141, 162, 182, 200.

As for Wallis, he defined air as 'the *Aggregate of both these*, or whatever else makes up that *Heterogeneous Fluid* wherein we breath[e], commonly called *Air*; the purer part of which is Mr. *Hobbes's Air*; and the feculent of it is Monsieur *Hugens's Air*'. In short, Wallis, as well as the other philosophers cited by him, treated the atmosphere as a mixture.

This kind of theory was widespread among seventeenth-century philosophers. In 1665, for example, Robert Hooke held that air was a collection of vapours and other light bodies suspended in the æther (refined matter), the latter remaining stagnant until forcibly moved.[73] In 1669, Boyle, who distinguished air from æther, claimed that air was a mixture of three kinds of particles: terrestrial effluvia, aqueous steams and aerial corpuscles—the last being air proper.[74] In 1673 William Brouncker supposed that the

> *purer* part of the Air to be of like nature with the *grosser* part ... and, though finer than the rest, so as to penetrate Glass, which the grosser will not (there being in all sorts of grains, some greater than others, and which will not pass so fine a Sieve;) yet of a Springy nature, as the *grosser* parts are: Which therefore acts, not by its *Weight* only, but by its *Spring*; and therefore when once entred, though in a small proportion, acts as effectually, as its first entrance, as if the whole incumbent Air had admission; its Spring being of a like tensure with that of the outward air....[75]

And in 1675 Newton seemed to agree, when he asserted that 'air is compounded of the phlegmatic body of air intermixed with various vapours and exhalations' and that æther is 'much of the same constitution with air, but far rarer, subtler, and more strongly elastic'.[76]

During the seventeenth century, however, nothing was known about the air's chemical properties, so that the philosophical debate concerned chiefly its physical properties—its elasticity. Descartes had explained elasticity by the random motion ('agitation') of particles of air caused by motion of the æther,

---

[73] Hooke, *Micrographia*, pp. 103–105. Hooke's æther is the same as Hobbes's, i.e., it is the passive principle, the active principle being heat; see Kassler, *Inner Music*, pp. 69–70, 157–158. For details of Hooke's mixture theory of the atmosphere, see Centore, *Robert Hooke's Contributions to Mechanics*, p. 61; Lysaght, 'Hooke's Theory of Combustion', pp. 94–99.

[74] Boyle, 'Certain Physiological Essays...' [2d edn. 1669], *The Works*, vol. 1, pp. 382, 393–394, 403.

[75] Wallis, 'An Extract of Letters', p. 5165. Prior to 1660 Brouncker made 'some trials' on the nature of air; see Boyle, 'New Experiments...' [2d edn., 1662], *The Works*, vol. 1, p. 160. who wrote that from Brouncker's 'further enquiries into this matter, if his occasions will allow him to make them, the curious may well hope for something very accurate'.

[76] Newton, *Papers & Letters on Natural Philosophy*, pp. 179–180; for ingredients in the mixture, see Newton, *Mathematical Principles*, e.g., pp. 383, 530, 547.

which pervades and separates them; but he conceived the invisible particles constituting the air as inelastically hard,[77] whereas Charleton, following Gassendi, explained the elasticity of the air by the flexibility of its invisible stationary particles.[78] In his *New Experiments, physico-mechanicall, touching the Spring of the Air*, published in 1660, Boyle introduced a new hydrostatic hypothesis, when he argued that the invisible aerial particles are springy and push against the weight (pressure) of the atmosphere.[79] Note that Boyle's hypothesis has two parts—spring *and* weight. Some twenty years previously, Evangelista Torricelli, a member of the Accademia del Cimento, had speculated that the earth's atmosphere had weight and that the weight was great enough to push mercury up in a yard-long tube. In 1643 a fellow academician set up an experimental test by taking a glass tube about a yard long, an inch wide and closed at one end. He filled the tube to the brim with mercury, thus forcing out the air in the tube. Then, covering the open end tightly with a finger, he turned the tube upside down, lowered the tube into an open dish of mercury until the tube was well below the surface of mercury in the dish, and lifted his finger from the mouth of the tube. When he did this, the mercury column, that had filled the tube, dropped to a height of 30 inches above the mercury in the dish and remained there, leaving a six-inch space at the top of the tube. This space, assumed to be empty, was later called the 'Torricellian vacuity'. Subsequently, Torricelli improved the experiment, and this was the origin of the barometer.

In 1654 Charleton provided the first published account of this experiment in England,[80] thereby providing a stimulus to the debate about the atmosphere, because the experiment 'raised questions of the highest interest for the interpretation of natural events'.[81] At first, the debate concerned the space above the mercury—for some philosophers, the space was empty, for others, it was filled with vapour or with æther. But from the 1660s, when Boyle presented his more developed hypothesis, the debate began to focus on whether

---

[77] See the summary from Descartes, *Principia philosophiæ* (1644) in Hesse, 'Hooke's Vibration Theory', p. 436.

[78] As Dijksterhuis, *The Mechanization of the World Picture*, p. 456, pointed out, a property possessed by the air as a whole is thus explained by assigning it unchanged to the particles constituting the whole; but 'this is no more enlightening than the method of making a quality responsible for it'.

[79] Boyle, 'New Experiments...' [2d edn. 1662], *The Works*, vol. 1, pp. 1–117 (1660), 118–242 (the 1662 additions).

[80] See Charleton, *Physiologia*, p. 348: 'the late famous Experiment of introducing a Vacuum in a Tube, by Water or Quick-silver, invented by *Torricellius*'; see also Webster, 'The Discovery of Boyle's Law', p. 457.

[81] Cromartie, *Sir Matthew Hale*, p. 209. For the Torricellian experiment, see Middleton, *The History of the Barometer*, pp. 19–32, and *The Experimenters*, pp. 105–109.

or not the air had a 'spring'. In 1673 the chief justice of the King's Bench, Matthew Hale, entered the fray, when, in a sequence of books, he argued that the space above the mercury was a vapour dissolved in the metal. His argument was based on the supposition that matter could condense and rarefy, whereas Gassendi and Descartes had rejected rarefaction, the former by conceiving it as an effect of a greater dilution of matter with the vacuum and the latter by conceiving the rarefying presence as æther. Hale's target was the atomist philosophy as reinterpreted in different ways by these two philosophers; but he also had Boyle in his sights and, particularly, Boyle's hydrostatic hypothesis, which Hale seems to have misunderstood.[82]

It is possible that Boyle himself contributed to Hale's misunderstanding, for, although his chief purpose in 1660 had been to demonstrate the existence of the elasticity of the air and to describe the phenomena to which it gives rise, he also proposed two different hypotheses—Cartesian and Gassendian—that might account for the cause of the 'spring of the air'.[83] Moreover, his presentation of the Cartesian hypothesis, which he repeated in 1662,[84] deviated from Descartes' own statement in two important ways. First, Boyle, but not Descartes, made the aerial particles flexible; second, Boyle, but not Descartes, made the aerial particles rotate. Mary Hesse has argued that Boyle's presentation was not his own but derived from Hooke. She supported her argument by citing correspondence to and from Huygens, who in 1662 had objected to Boyle's presentation on the grounds that air so constituted would lose its fluidity. Subsequently, Huygens was sent a letter that Hooke had written to Boyle in July 1662, defending his 'Epicurean' hypothesis and explaining, among other things, that fluidity is preserved, because the particles are able to slide by each other. Huygens then expressed his approval of the hypothesis, stating in December 1662 that it was 'well explained' by Hooke.[85]

By 1665, Hooke had changed his mind, for he had a clear grasp that the particles of 'springy bodies'—fluid as well as solid—vibrate, not rotate. Hence, he was beginning to develop Hobbes's kinematic variant of the continuum hypothesis, with its model of impact as an elastic process. On 22 June 1676 he utilised this variant in a lecture to explain the phenomena of hearing—sound

[82] For Descartes and revived atomism, see Dijksterhuis, *The Mechanization of the World Picture*, p. 417; for a summary of the English debate, as well as the secondary sources concerning it, see Cromartie, *Sir Matthew Hale*, pp. 209–217.

[83] Boyle, 'New Experiments...' [2d edn. 1662], *The Works*, vol. 1, pp. 11–12.

[84] The repetition appears in the additions to the first edition; see Boyle, 'New Experiments...' [2d edn. 1662], *The Works*, vol. 1, pp. 178–179.

[85] See Hesse, 'Hooke's Vibration Theory'; for the relevant correspondence, see Huygens, *Œuvres complètes*, vol. 4, pp. 171, 217, 221, 274.

and music.[86] But the main innovation in this lecture is Hooke's theory of hearing, according to which the eardrum, functioning on the analogue of a stretched string, regulates its tension so that its vibrations are 'tuned unison'—that is, keep time with incoming vibrations.[87] If the two sets of vibrations are in time, there is pleasure; but if they are out of time, there is pain, since, in the latter case, the eardrum is overstretched. In a brief conclusion, Hooke used this theory to suggest that if the slower or faster timings are within the range of human audibility, they operate on 'the affections & passions of men' not only as variations of pitch, duration ('time'), loudness or softness of tones but also as consecution of tones diversified by metre ('measure') and, most especially, agreement of tones sounded together in harmony.[88] With this suggestion, Hooke provided the first hint that a connecting link could be made between the sciences of mechanics and music.

THE 1677 *ESSAY*

As previously indicated and as shall appear subsequently, an actual link between the sciences of mechanics and music was first made in 1677, when a slim quarto volume of only thirty-five pages was published anonymously as *A Philosophical Essay of Musick directed to a Friend*, price 1 shilling stitched.[89] The author of the tract seems to have been unknown until 1740, when he was publicly identified as Francis North (1637–85).[90] Trained as a lawyer in the Middle Temple, London, he was called to the bar (1661) and bench (1668), becoming king's-counsel (1668), solicitor-general (1671), attorney-general (1673),[91] chief justice of the Common Pleas (1675), member of the Privy

[86] The lecture remained unpublished until transcribed by Gouk, 'The Role of Acoustics', pp. 597–605.

[87] See Kassler, *Inner Music*, pp. 131–133.

[88] Hooke in Gouk, 'The Role of Acoustics', p. 604–605.

[89] The price, but not the author's name, is given in Arber, *The Term Catalogues*, vol. 1, p. 274 (Easter term, 28 May 1677).

[90] See RN, *Life/FN*, pp. 249–250. The first *published* attribution would have been in the 1740 version, edited and mangled by RN's second son, Montagu. For RN's other references to the *Essay*, see *Notes of Me*, p.160; *Cursory Notes of Musicke*, p. 172; *NP 3*, pp. 62, 76; and Appendix A, pp. 179–200 *infra*.

[91] KING'S-COUNSEL: on the nomination of the lord chancellor, a barrister appointed counsel to the crown takes precedence over ordinary barristers. SOLICITOR-GENERAL: a law-officer, in rank next to the attorney-general, who takes the part of the state or crown in suits affecting the public interest. ATTORNEY-GENERAL: the first ministerial law-officer of the state or crown empowered to act in all cases in which the state is a party.

Francis North, 1st baron Guilford (1683)
[Engraved from an original portrait attributed to John Riley]

Figure 1.3

Council (1679) and, finally, lord keeper of the Great Seal to Charles II (1682) and to James II (1685).[92] In May 1671 he was knighted; and on 22 September 1683 he was raised to the peerage as baron Guilford [*Figure 1.3*].

North was introduced early to music in the household of his grandfather, where there were a number of resident musicians.[93] Indeed, as John Aubrey later remarked, 'I remember that he told me that he learnt it so young he could hardly remember it: it was so habitual with him'.[94] Afterwards, he was educated at King Edward VI free grammar school in Bury St. Edmunds; and on 8 June 1653 he entered St. John's College, Cambridge, where he continued to apply himself to music, becoming 'a most expert violist, and at length a composer of consorts'.[95] He also studied philosophy, old as well as new, assisted by Edmund Matthews.[96] During his more carefree days in London, he was 'a noted hunter of music meetings'[97] and a patron of some in the music profession.[98] And even after his professional responsibilities began to weigh

[92] He never was lord chancellor, as some secondary sources claim.

[93] The household musicians included Henry Loosemore, George Loosemore, John Jenkins and John Lilly; see Randall, *Gentle Flame*, pp. 83–84; Wilson, *Roger North*, p. 37 and Peter Millard in RN, *Notes of Me*, p. 276. For FN's copies of Jenkins' music, see Willets, 'Autograph Music by John Jenkins', and Holman, 'Suites by Jenkins rediscovered'.

[94] See Stephens (ed.), *Aubrey on Education*, p. 37, where Aubrey also describes FN as 'an admirable musician'. For Aubrey and FN, see Appendix B, pp. 201–202 *infra*.

[95] RN, *Life/FN*, p. 12 and pp. 14, 44, 248; see also RN, *Notes of Me*, p. 157, for FN's love of Italian vocal music—he had 'no voice, but exquisite skill'. In 1654, DN (1) presented FN with a 'new' viol, which implies he had learned to play the instrument prior to university; see Randall, *Gentle Flame*, p. 72. For a composition by FN included in one of the many surviving manuscripts of Jenkins' music, see (GB:Ob) MS North e.37.

[96] RN, *Life/FN*, p. 12. Edmund Matthews [Matthewe], B.A. (1637), M.A. (1640), B.D. (1647), was a fellow at Sidney College, Cambridge, where he matriculated (1632) and where he had been tutor to Charles North, eldest son and heir of DN (1).

[97] According to RN, *Life/FN*, p. 24, 'that diversion helped him to keep out of the ways of vice'.

[98] On 3 February 1672 FN recommended that the marshall wardens and assistants of the Corporation for Regulating the Art and Science of Music admit George Pigot [Piggot, Pigott] as clerk of the Corporation; see Ashbee (comp.), *Records of English Court Music*, vol. 5, p. 260, and *NP 4*, p. 42. Pigot had been a member of the Musick Society that gave concerts in the Old Jewry between 1659–65, two other members of which had been educated at the same grammar school as FN. John Playford, clerk to the Temple Church, published at least one book of songs for the Musick Society, so it is not without interest that FN owned an inscribed copy of Playford's, *Psalms and Hymns in solemn Musick* (London, 1671). See *NP 4*, p. 43; Wilson, *Roger North*, p. 36n.; and Spink, 'The Old Jewry "Music Society"'.

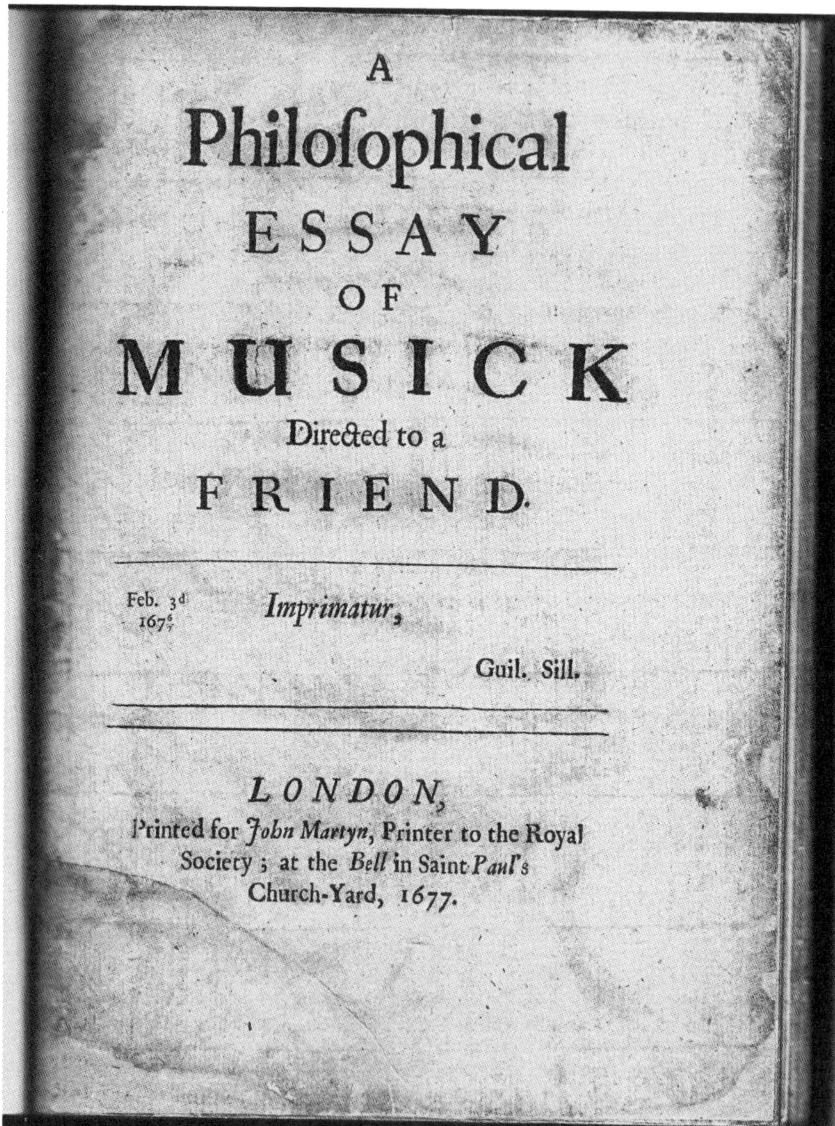

A
Philoſophical
ESSAY
OF
MUSICK
Directed to a
FRIEND.

Feb. 3d 167 6/7 *Imprimatur*,

Guil. Sill.

LONDON,
Printed for *John Martyn*, Printer to the Royal Society; at the *Bell* in Saint *Paul's* Church-Yard, 1677.

Title page: *A Philosophical Essay of Musick*

Figure 1.4

heavily, he continued to find solace in music until the politics of faction and corrupt practices destroyed his health and brought about his early death at the age of forty-eight.[99]

I. THE GENESIS OF THE *ESSAY*

The circumstances that led to publication of North's *Essay* are shrouded in mystery.[100] The colophon indicates that its printer, John Martyn, was also printer to the Royal Society [*Figure 1.4*]. But, according to that Society's regulations, the author of a book issued under its auspices had to be a fellow; and for Martyn to print such a book, the Society's licence was required, and a member of the Society's council was to advise the printer as to 'the correctness of the edition, the number of copies printed, the form or volume, the goodness of the paper, character, figures and diagrams, the price at which books are to be sold'.[101] North, however, was not a fellow, although a section in the Society's 1663 charter allowed non-fellows to be present at meetings. But on 15 October 1674 the council minuted its intention to repeal that section; and up to that date there is no evidence in the Society's records that North attended any of the meetings.[102]

---

[99] During the period when he was lord keeper of the Great Seal, his 'greatest releif of mind was musick', so that, for him, 'musica mentis erat medicina mestæ'; see RN, *Life/FN*, pp. 121, 222. During this period, too, Henry Purcell brought some of his music to FN's residence (then in Queen Street), performing it there with RN and 'another violin', perhaps Nicola Matteis; see Wilson, *Roger North*, pp. 47–48. Except for RN, *Life/FN*, there is no biography that addresses the problems of reception history or that provides a balanced account of the life of a man who lived during a period of moral decay in politics, for which see Birrell, 'Roger North and Political Morality'.

[100] H.W. Turnbull claimed, without giving any evidence, that FN wrote the *Essay* while still at school; see Newton, *The Correspondence*, vol. 2 (edited by Turnbull), p. 207. But Turnbull's claim seems questionable, because, as indicated in Chapter 2 *infra*, the *Essay* contains insights that are in advance of what was known during FN's student days.

[101] According to Rostenberg, *The Library of Robert Hooke*, p. 15, Martyn was required to supply two copies to the king; one copy each to the lord chancellor and to the president and secretaries of the Royal Society; and two copies to the library of the Society (which does not now possess copies of FN's *Essay*). According to the registers of the Worshipful Company of Stationers, it was Sarah Martyn who in 1683 entered many of her husband's books at Stationers Hall, including FN's *Essay*; see Eyre (ed.), *A Transcript*, vol. 3, pp. 181–190.

[102] Lack of evidence is not decisive, however, because on 23 May 1667 Margaret Cavendish, duchess of Newcastle, had 'intimated' her desire to be present at some of the meetings, so as to be 'entertained' by some experiments; and it was minuted that such instances were not to be inserted in the 'history' of the Society; see Birch, *The History of the Royal Society*, vol. 2, pp. 175–176, and vol. 3, p. 137; see also Pepys, *The Diary*, vol. 8, p. 243 and n.1.

In that year a new declaration for regulating the Society was drafted; and one of the fellows, Robert Southwell, read the declaration to the 'Lord Keeper'.[103] On 9 November 1674, Southwell thanked the council for having procured for the Society 'from Mr Attorney Generall that Legal forme of Subscription' for weekly payments of fellows; and he also expressed his desire that the council assure both the 'Lord Keeper' and 'Mr Attorney Generall' of the council's 'deep sence they had of their favour to the Society'.[104] 'Mr Attorney Generall' was none other than Francis North, who held this post from 12 November 1673 until 25 January 1675, when he became chief justice of the Common Pleas. On 22 May 1675, North wrote to the Society's secretary concerning the swim bladder of fish (its hydrostatic organ).[105] This letter seems to be missing from the Society's original records, but it was included in the Society's letter book copy as 'A Letter containing a conjecture concerning the Bladders of Air that are found in fishes communicated by an anonymus [sic] to M. Oldenburg and by him to the Royal Society May 22, '75'.[106] Two days later, the letter was printed in the Society's *Transactions*, with an added experiment of Robert Boyle to illustrate North's conjecture.[107]

Authorship of the letter was first revealed in the memorials of Roger North, who reported that his brother, Francis, was prompted to consider some problems in hydrostatics as a consequence of reading a book by Matthew Hale.[108] As noted previously, Hale disliked Boyle's doctrine of hydrostatic

---

103 Heneage Finch, bt. (1660), was successively attorney-general (1670), lord keeper of the seals (1673), baron Finch and lord chancellor (1674) and 1st earl of Nottingham (1681). Although a zealous supporter of court policy, he was independent as a judge. See RN, *Life/FN* pp. 19, 40, 42–43, 46, 105, 108–109, 111, 175, 179–180, 182–183, 208, 212, 231, 273, 311, 330–331, 358, 373, 428, 456–458. His brother, John, an original fellow of the Royal Society, took a prominent part in drawing up the Society's 1662 and 1663 charters; see Purver, *The Royal Society*, pp. 17–18. A son, Edward, composed music and wrote a thorough-bass 'grammar'; see Kassler, *The Science of Music*, vol. 1, pp. 352–354, and Herissone, *Music Theory*.

104 See Birch, *The History of the Royal Society*, vol. 3, p. 139–141.

105 Henry Oldenburg, the Society's first secretary, was born at Bremen and lived in England 1640–48, returning again in 1653. During studies in Oxford, 1656–57, he became acquainted with Robert Boyle and between 1657–60 travelled as tutor to Boyle's nephew, Richard Jones, afterwards lord Ranelagh. For his association with Joseph Williamson as one of the licensers of the press (6 February 1675 and 29 April 1676), see Kitchin, *Sir Roger L'Estrange*, pp. 195–196, 199.

106 (GB:Lrs) LBC, vol. 7, 358–360. This letter was not included in Oldenburg, *The Correspondence*.

107 See FN, 'A Conjecture'. For an approval of the conjecture, see Ray, 'A Letter'.

108 RN transcribed the letter and recorded that it had been sent to Oldenburg without FN's name; see RN, *Life/FN*, pp. 244–245. The book in question was Matthew Hale, *An Essay touching the Gravitation or Non-gravitation of Fluids* (London, 1673, 2d edn. 1675). For a

pressure, so North 'fell upon' the consideration of swim bladders, 'what effect they had, and by what means'.[109] He concluded that contracting and dilating the bladder caused the fish to rise, sink or rest in the water, without any use of its fins. And, as was his practice, he tried out his 'conjecture' with one of his 'friends'—Roger North himself. But the latter alleged against it that there is no occasion when a fish does not use its fins, just as there is no occasion when people do not use their muscles 'to stand, or sit, upright', although that action—the tonic motion of the muscles—is altogether involuntary.[110] But Francis North did not yield his position; and at length, 'it was agreed, that each party should draw up their reasons and send them to Oldenburg', 'as to a publik notice'.[111]

This was not the only occasion that a disagreement between the two brothers resulted in letters to the Royal Society, since Roger North reported that he and Francis once had

> a hot dispute about the barometer, and the reasons of its keeping time with the weather. And agreed to putt our sentiments in wrighting and send them to Oldenbury [sic] of the Royall Society to be printed. And he returned them all, which made my brother Francis laugh, and say his fared ill for being in such company. I thought then and doe still that I was in the right, and have set forth the matter at larg in my discours of the barometer, which had its start from this controversie. My brother F[rancis] went on the common notion of vapours, but made them heavyer when low, and ready to raine, then when high; as all gravity is less farr from then neer the surface of the earth. I battelled these vapours, and persisted that the common air was water, or vapour, and that heats, and colds intermixing swealled, or shrank the atmosphear, by producing the effects of fair weather or rain; and that influenct the mercury.[112]

After 1675, however, there is no mention of Francis North in the Society's records.

Fortunately, additional information occurs in the daily diary kept by the Society's curator of experiments, Robert Hooke, who, on 10 November 1676,

---

discussion of this work, as well as two related works by Hale, see Cromartie, *Matthew Hale*, pp. 209–217. As colleagues in the law, there was some antipathy between FN and Hale, for which see Peter Millard's comments in RN, *Notes of Me*, pp. 284–285.

[109] Against Boyle's doctrine, Hale promoted the 'funicular' hypothesis of the English Jesuit, Francis Line (or Hall); see Reilly, 'Francis Line, Peripatetic', pp. 231–234.

[110] For RN's explanation of tonic motion, see Kassler, *Inner Music*, pp. 195–196; for a later, related dispute with his brother, DN (2), see Grassby, *The English Gentleman in Trade*, p. 335.

[111] RN, *Life/FN*, p. 244. RN's letter is not part of the list of correspondence compiled by Korsten, *Roger North*, pp. 232–252, so that his and FN's letters may no longer be extant.

[112] RN, *Notes of Me*, p. 201. This dispute, and the papers resulting, would have taken place before the death of Oldenburg in September 1677; see also Appendix B, p. 209 *infra*.

recorded: 'Received paper about motion from Lord North among letters. Walkd with Sir Ch. Wren in St. James Park. About motion.'[113] The Lord North in question was Francis North's father, Dudley, 4th baron North,[114] to whom Hooke replied on 12 November 1676, recording in his diary: 'Wrote answer to motion query.' Then Hooke seems to have received a second letter from a 'Mr. North', for on 26 November 1676, the diary entry reads: 'Wrote answer to Mr. North of motion.'[115] Note that, in the first entry, Hooke indicated the paper was *from* Lord North, not *by* him, so that, taking into consideration the last entry, it seems that all three memoranda relate to Mr. North. If this Mr. North were Francis, Hooke would have referred to him as 'Lord Chief Justice North', for he observed the niceties of address even in his diary.[116] Nevertheless, that the query about motion was related to the *Essay* seems certain from three further entries in Hooke's diary for 15, 19 and 23 February 1677, when he visited John Martyn, and recorded: 'Saw at Martins tract about Musick.' 'At Martins the Musick sheet.' 'From Martin book of musick 1sh. not paid.'[117] From these brief memoranda, we may conclude that the *Essay* was published prior to 15 February 1677. In the month following, a synopsis was published in the Royal Society's *Transactions*.[118] About this time also, John North, another brother of Francis, sought comments on the *Essay* from Isaac Newton, who, on 21 April 1677, replied by letter, because, as indicated by the postmark, John North was then in London.[119]

As Hooke's memoranda indicate, Martyn was a printer as well as a publisher and bookseller. It is likely, therefore, that Martyn printed the *Essay* not for the Royal Society but on behalf of one of the Norths, possibly John, who had been involved with arrangements for publishing some of Lord North's writings, one of which was printed by Martyn.[120] This inference is supported

---

[113] Hooke, *The Diary*, p. 256.

[114] For biographical details of DN (1), see Abbreviations, pp. xi–xii *supra*.

[115] Hooke, *The Diary*, pp. 256, 259.

[116] For his reference to FN as 'Lord Chief Justice', see Hooke, *The Diary*, p. 341.

[117] Hooke, *The Diary*, pp. 274–275. For 'the Musick sheet', see Appendix A, p. 137 (*Figure 1.A.*) *infra*.

[118] See Appendix A, pp. 171–174 *infra*.

[119] See Appendix A, pp. 175–178 *infra*. The letter's address indicates that it was to be left at 'Mr Pawley's' at the Bible in Chancery Lane, London. This was the address of Robert Pawlett, a bookseller with whom JN had previous dealings; see below n.120. The postmark is 'AP' over '23' in a circle, meaning that the letter was received in London on 23 April. For biographical details of JN, see Abbreviations, p. xiii *supra* and further *infra*.

[120] In May 1669 JN visited Martyn to make arrangements for printing his father's *Observations and Advices Œconomical*, which was licensed on 28 June 1669 and published anonymously, price 1 shilling. Hooke owned a copy of this book; see Rostenberg, *The Library*

by the licence authorising the printing of the tract, which was granted on 3 February 1677 by William Sill ('Guil. Sill.') [see *Figure 1.4*].[121] During the 1670s Sill, a clergyman and chaplain to the bishop of London, signed imprimaturs on behalf of that bishop, Henry Compton, who from 1675 was also dean of the Chapel Royal.[122] In the latter capacity, Compton was part of the clergy attached to the court at a time when John North was clerk of the closet to Charles II, an office that included attending the king during divine service, waiting on him in his private 'closet' and resolving doubts about spiritual matters.[123] It would seem, therefore, that this was one of several periods when John North resided in London.

Although not all of John North's visits to London can be dated with certainty, he seems to have resided there 'a little' with his 'best brother', during which time he became acquainted with some of the gentlemen of the society of the Temple.[124] This must have been after the death of Geoffrey Palmer in 1670, when Francis North moved into Palmer's chambers in the Middle Temple and

---

of Robert Hooke, p. 194 (item 159). In the following year arrangements were made with Pawlett in Chancery Lane for printing another tract of DN (1), *A Narrative of Some Passages in or relating to the Long Parliament*. Pawlett took the manuscript to the licenser, Roger L'Estrange, whose objections to some parts of it were answered by JN. When L'Estrange 'relented insufficiently', JN sent the manuscript to Samuel Parker, chaplain to the archbishop of Canterbury, who licensed it on 12 November 1670. Like the earlier work, the tract was published anonymously, price 1 shilling. See Randall, *Gentle Flame*, pp. 81, 86–87.

[121] William Sill graduated B.A. and M.A. from Trinity College, Cambridge. Subsequently, he was vicar of Addingham, Cumberland (1668–73); prebendary of Carlisle (1668–81); vicar of Bromfield and of Torpenhow (1673–81); prebendary of St. Paul's (1678–81); archdeacon of Colchester (1678–81); rector of St. Augustine's, Watling Street, London (1680–87); and prebendary of Westminster (1681–87).

[122] Henry Compton was translated from Oxford to London in 1675, when he also became dean of the Chapel Royal and in 1676 a privy councillor. His advancements in the church, 1673–77, came through family influence and by favour of the lord treasurer, Thomas Osborne, 1st earl of Danby; but Compton's hopes for the see of Canterbury were frustrated in 1677 by the duke of York.

[123] The clergy attached to the court consisted of the dean of the chapel; the sub-dean; twelve priests, who, with twenty lay clerks of the chapel, conducted the services in the Chapel Royal; forty-eight chaplains in ordinary, who usually were doctors of divinity; a number of extraordinary chaplains; a lord high almoner and sub-almoner; and the clerk of the closet. For JN's duties, see Peter Millard's note in RN, *Life/JN*, pp. 184–185, who also points out that the date of appointment is uncertain, the evidence placing it sometime between September 1674 and September 1676.

[124] RN, *Life/JN*, p. 112. His acquaintances included (1) in the Middle Temple, two brothers of Sidney Godolphin, who in 1675 married Margaret Blagge, a third cousin of JN and a close friend of John Evelyn; (2) in the Inner Temple, John King, a king's-counsel and attorney-general to the duke of York, who was knighted in 1674, and William Longueville, conveyancer, patron of Samuel Butler and friend of FN. For a brief character of Longueville, see RN, 'Letters', pp. 237–239.

when Roger North also resided there.[125] After 1673, when John North became a prebendary of Westminster, he had access to 'an house and accommodations for living in town', facilities which he seems to have used from time to time.[126] Hence, the Mr. North referred to by Hooke could have been either John or Roger North.[127]

II. THE *ESSAY'S* 'FRIEND'

So far, these details, however scanty, relate to the genesis of the *Essay*. What, then, about the friend to whom it was directed? In the epistle that introduces the *Essay* (pp. iii–iv), Francis North not only indicates that he 'often tyred' this friend with 'discourses of several parts of musick' but also described the friend as 'so great a philosopher and musician as you are'. With one exception, to be mentioned later, these are the only overt clues that might help to identify the friend. On a previous, but—as now is apparent—erroneous supposition that the Royal Society was involved in the genesis of the *Essay*, I conjectured that a possible candidate might be William Brouncker, 2d viscount Brouncker of Castle Lyons in the Irish peerage, who held the office of president of the Royal Society from its first incorporation to 30 November 1677.[128]

Although Brouncker was involved in investigations relating to the atmosphere, as well as experiments on the recoil of guns, he is known today for his work on mathematics, an interest that is reflected in his only published work on music—the *Animadversions* that was included, with a separate

---

[125] Geoffrey Palmer, a barrister of the Middle Temple (1623), was created 1st baronet and nominated attorney-general at the Restoration. One of his sons, Edward, was a close friend of FN. RN took up residence with FN 'first while Sir Jeffry Palmer was but just alive'; afterwards he moved with FN to Palmer's chambers, where FN's 'greatest content was, that he had means to accomodate his brother of Cambridg [JN], when he thought fitt to come to London, so as to have him so near, that he could at any time goe from his buissness to him, and returne to it againe when he pleased'; see RN, *Life/FN*, pp. 39 and 44.

[126] RN, *Life/JN*, p. 124. He was installed on 11 January 1673, two days after presentation by the king.

[127] These are not the only candidates, of course, since Mr. North could have been FN's favourite clerk, Robert North; see RN, *Life/FN*, pp. 44, 187–189, 365–366, and Randall, *Gentle Flame*, p. 78. There is the further complication that there is no mention in the published Cambridge records of JN's doctorate of divinity. In his notes to RN, *Life/JN*, pp. 184–185, Peter Millard states that since the ceremony awarding the degree took place when Isaac Barrow was vice-chancellor, 'it must have been between 1675 and 1676'. When Evelyn, *The Diary*, vol. 4, pp. 87, 97, heard JN preach on 24 March 1676, he referred to him as 'Dr.'; but when, later on, he met JN for the first time, he referred to him as 'Mr. North'.

[128] Chan and Kassler, 'Francis North', p. 52.

titlepage, in the English translation of Descartes' *Compendium musicæ.*[129] Although Brouncker was concerned chiefly with various means of dividing the scale, he promised to expand on his new theory in '*an entire and particular* Tract', a promise that seems not to have been fulfilled.[130] Instead, during his tenure as president of the Royal Society, he encouraged investigations by others that tended to the advancement of music, both theoretical and practical,[131] whereas in a private capacity he was a hunter after, or a host of, music meetings.[132]

Francis North seems to have owned a copy of one of Brouncker's mathematical papers, written in reply to Hobbes and read on 4 September 1661 at a meeting of the then unincorporated Royal Society. Hobbes had submitted to Charles II a paper containing a geometrical proposition to find two mean proportionals between two straight lines; and the king passed on this paper to the Royal Society. Before it was made public, Brouncker produced a refutation.[133] How the paper, or a copy of it, came into the hands of Francis North is unknown. But, to date, this is the only evidence that Brouncker and North might have been acquainted, even though contact between the two men could have been made through any one of a number of Francis North's friends or relations who were fellows of the Royal Society.

A second candidate was proposed by John Wilson, who wrote that the unnamed friend 'must surely have been' Roger North.[134] During the seventeenth century it was not uncommon to apply the word 'friend' to a near relation. And in the biographies of three of his brothers, Roger North often

---

[129] Although the appended *Animadversions* was published anonymously by a 'Person of Honour', attribution of authorship was made by Brouncker's friend; see Charleton, *Physiologia*, p. 226. According to some, e.g., Chenette, *Music Theory*, p. 404, Brouncker was the translator of Descartes' tract. But in the preface to the reader, the 'Stationer' clearly states that he is the translator; and since he makes an error in describing Brouncker's system, the 'Stationer' and the 'Person of Honour' (Brouncker) must be distinct persons. For information about the 'Stationer'—one or the other of the two booksellers on the colophon of Descartes, *Excellent Compendium*—see Plomer, *A Dictionary of the Booksellers and Printers*, pp. 95, 134–135.

[130] Brouncker in Descartes, *Excellent Compendium*, p. 91.

[131] For a summary of these investigations, see Gouk, 'Acoustics in the Early Royal Society'.

[132] For Brouncker as philosopher of music and as music lover, see Pepys, *The Diary*, vol. 9, p. 147 and vol. 8, pp. 25, 54–57, 64–65.

[133] For the history of this paper internal to the Royal Society, see Huygens, *Œuvres complètes*, vol. 3, pp. 336, 368–370, 383–385; for Brouncker's refutation, see (GB:Lrs) RBO, RBC 1.104; for FN's summary, see RN, *Life/FN*, pp. 166–167.

[134] Wilson, *Roger North*, p. 41.

appears in just this guise, as 'a friend' or 'a young friend'.[135] Nevertheless, Wilson's candidate seems improbable for two reasons. First, Newton's letter, commenting on the *Essay*, came to the attention of Roger North only in the 1720s.[136] Yet, if he were the friend, surely he would have known about this letter. Second, the introductory epistle to the *Essay* is written in a style that suggests the friend is an equal. But Roger North was fourteen years younger than Francis, who acted toward him as a patron, mentor and, after 1677, even a father.[137] Between 1669 and 1674—the period when Francis North was king's-counsel, solicitor-general and attorney-general, Roger North was a student of the law. In January 1675, when Francis North became chief justice of the Common Pleas, Roger North commenced legal practice, which included riding the circuit with his brother. But even then the relationship was unequal, as Roger North himself hinted when referring to the 'many mortifications, by little contempts my brother [Francis], sometimes in jest, and often in earnest would put upon me'.[138]

There is no doubt that the two brothers were passionate about music; and there is evidence of lively debates about those parts of its philosophy concerned with the propagation of sound, the constitution of the air as the medium of sound and the initiation of vibrations in an organ pipe. But it was not until some twenty years after the *Essay* was published that Roger North began to produce his numerous drafts, essays and treatises on the philosophy of music.[139] Hence, it seems we need to look elsewhere for another candidate—perhaps John North.[140] This brother acted as a spark of, as well as a moderator in, the debates between Francis and Roger North, since, as the latter wrote:

> Our conversation was much upon musick, and philosofy. Our brother, John, who was a Cantabrigian, and very ingenious as well as learned, gave much life to our enterteinements, for he was excellent at sparring

---

[135] See, e.g., RN, *Life/FN*, p. 121: 'His [FN's] greatest releif of mind was musick, of which he comonly took a relish at his going to bed, for which end he had an harpsicord at his bedchamber door, which a freind [RN] touched to his voice.'

[136] See Chapter 3, p. 102 *infra*.

[137] See Peter Millard's 'Introduction' to RN, *Notes of Me*, pp. 8–9, 11.

[138] RN, *Notes of Me*, p. 160, also indicates that FN had 'some what of humour that way, of raising his owne, by depressing other's caracters'. For an example, see RN, *Life/FN*, ('Of frequent disputes in philosofye') pp. 261–266, where RN reproduces a paper 'about gravity', undoubtedly by himself, with FN's response.

[139] See Kassler, 'Roger North', p. 55.

[140] In RN, *Life/JN*, p. 134, FN is described as JN's 'best friend'; and in a letter from JN to FN (GB:Ob) MS North c.5: f. 71, JN writes: 'my Dearest Brother and cheifest Freind'.

> doubdts, and reserved himself to be a moderater, of our disputes which would be sometimes very warm.[141]

Like his brothers, John North seems to have been introduced to music when young. But, unlike his brothers, music never became

> a favourite diversion or manual exercise to rest his mind a little, which he held bent with continual thinking. His parents, who were much addicted to music, recommended that to him for a diversion, and particularly the noble organ, as the fullest and not only a complete solitary consort, but most proper for an ecclesiastic. And indeed if study had not had the upper hand of all his intendments, he must of course have taken up that way, being *quasi ex traduce*, for after the care of prayers and meals, nothing was more constant and solemn than music was in that family. He [John North] was sensible the advice was very good, and accordingly got a small organ into his chamber at Jesus College, and suffered himself to be taught a lesson or two, which he practised over when he had a mind to be unbent.[142]

Unfortunately, a person occupying the room below was annoyed at the noise from the 'bellows knocking on the floor and the hum of the pipes'. Hence, John North 'made no manner of advance', for there could be no regular practice that leads not only to musical skill but also to pleasure in music, which, 'like that of books', is 'never true and good unless easily and familiarly read or performed'.[143]

How, then, could Francis North describe John North as 'so great a philosopher and musician as you are'? The answer, perhaps, is that the adjective 'so great' was part of the customary florid style used in a letter which functions somewhat like an epistolary dedication; for the dedicatee, if John North, had 'great' connections at court, as well as a 'great' reputation at the university—indeed, Thomas Gale wrote of him as a most illustrious man 'who equals the glory of his family (which is certainly great) in virtue and learning'.[144] Certainly, his rise at the university was rapid, for he was made a fellow of Jesus College by royal mandate (1666); and after becoming a deacon

---

141 RN, *Notes of Me*, p. 201.

142 RN, *Life/JN*, pp. 109; 143. For JN's 'unmusical' nature, see Kassler, *Inner Music*, pp. 201–205.

143 RN, *Life/JN*, p. 109.

144 See RN, *Life/JN*, p. 128 and n.58:2. Thomas Gale was educated at Westminster School and Trinity College, Cambridge, where he was professor of Greek (1666–72). On leaving the professorship, he became high master of St. Paul's School (1672–97) and dean of York (1697–1702). From 1677 he was an active member of the Royal Society and is frequently mentioned in Hooke, *The Diary*.

(1669) and priest (1670), he succeeded Gale as regius professor of Greek (1672–74), resigning, it seems, when he became clerk of the closet to Charles II.

At the time of his appointment to the professorship, John North also changed colleges, migrating from Jesus to Trinity. According to Roger North, there were 'many reasons for his choice of that college':

> as that he valued much the company of that society where resided many excellent persons, and he perceived more of the humane and polite in that, than in the lesser colleges. But above all, the leading card was the value he had for the more than thrice excellent master, Dr. Barrow. He had long before contracted a familiar acquaintance, I may say friendship, with him, and they used each other in a most delightful communication of thoughts.[145]

Mathematics was one of the subjects of their discussion, so that John North kept Barrow abreast of the 'pains' he had taken in an attempt 'to prove an arc of a circle equal to a straight line (in order to square the circle)'.[146] But the two men shared much more in common than mathematics, since both were royalists; both were Anglican priests with a leaning towards the Arminianism of the Cambridge Platonists;[147] both were trained in classical learning; both had studied the philosophy of Descartes;[148] both were acquainted with, and

---

[145] RN, *Life/JN*, p. 119. Isaac Barrow was admitted first to Peterhouse and then Trinity College, Cambridge (1643); B.A. (1648), fellow (1649), M.A. (1652) and, by royal mandate, D.D. (1670). By Cromwellian mandate he was unable to succeed his tutor, James Duport, as professor of Greek, possibly because of his Arminian views; but at the Restoration he was appointed Cambridge professor of Greek (1660–63) and then Lucasian professor of mathematics (1663–69). From about 1662–63 he also was professor of geometry at Gresham College, London, at the same time acting as a locum for the professor of astronomy, either Lawrence Rooke or his successor, Christopher Wren. In 1673 he succeeded John Pearson as master of Trinity College (Pearson had been master of Jesus when JN was admitted there). In 1675 he was university vice chancellor. Although an original fellow of the Royal Society, Barrow seems not to have been an active member.

[146] RN, *Life/JN*, p. 129. From a hint in one of Barrow's Lucasian lectures, it seems this conversation may have taken place about 1664–65; see Barrow, *The Usefulness of Mathematical Learning*, pp. 201–202.

[147] For JN's Arminianism, see RN, *Life/JN*, p. 135; for the Dutch theologian, Jacobus Arminius and for general background, see Colie, *Light and Enlightenment*.

[148] For Barrow's study of the Cartesian philosophy, see his 1652 M.A. thesis, *Cartesiana hypothesis de materia et motu haud satisfacit præcipuis naturæ phænomenis*; see also his Lucasian lectures, *The Usefulness of Mathematical Learning*. For JN's study of Cartesianism, see RN, *Life/JN*, pp. 130, 136; see also Korsten, *Roger North*, p. 68.

admirers of, Newton;[149] both were masters of Trinity College, John North succeeding Barrow;[150] and both died young, Barrow aged forty seven and John North aged thirty eight.[151]

Here my case rests for John North as the friend to whom the *Essay* may have been directed, since until further information comes to light, the identity of that friend remains uncertain. But my case is weakened by a single sentence in the introductory letter, where Francis North writes (p. iv): 'I hope for this fruit of my pains, that you will make such experiments as your curiosity will suggest whilst you read this, and let me know the success of them.'[152] Since John North seems to have had no inclination to experiment, there is one other person who deserves consideration as the friend. This is Robert Hooke, who had a number of associates in common with North.[153] During the period leading up to 1677, for example, Hooke was in almost daily contact with one of the fellows of the Royal Society, John Hoskins, to whom Roger North gave 'an advanc't post' among his brother's 'ingenious' friends and acquaintances.[154] The two men became acquainted when Hoskins 'began to be eminent in practise' and when Francis North 'had some aspect towards the law'.[155] Although Hoskins was to become 'an accomplisht master in Chancery' and although North was to support his practice by sending 'references' to him, their friendship seems to have been cemented early by a common interest in philosophy, for they 'often gott to supper together, and the conversation was intirely upon experimentall philosofye, astronomy, etc.'.[156] Since there is little

[149] Newton matriculated at Trinity College in the same year that JN matriculated at Jesus (1661), and the two were in residence at Trinity in 1672. The contact between Newton and Barrow seems to have begun only after 1667, according to Whiteside, 'Isaac Barrow', p. 475.

[150] For the unhappy consequences of JN's mastership, see RN, *Life/JN*, pp. 144–161.

[151] For Barrow's last illness and place of death 'in London, in a prebend's house', see RN, *Life/JN*, pp. 119–120; see also Dick (ed.), *Aubrey's Brief Lives*, p. 183.

[152] For someone who did make the experiments, see Chapter 3, p. 89 *infra*.

[153] Two associates were related to the North family. One was Ralph Montagu; see RN, *Life/JN*, appendix iv. The other was Richard Edgcumbe, a fellow of the Royal Society (1676), whose wife was related to the Norths. Hooke designed Montagu House in Bloomsbury, and he also drew up a draught for improvements to one of the houses owned by Edgcumbe; see Batten, *The Architecture of Robert Hooke*, pp. 93–96, 110–111, and 'Espinasse, *Robert Hooke*, pp. 97–99, 105, 132, 150, 177. Work on both designs commenced prior to 1677.

[154] RN, *Life/FN*, p. 242. For Hoskins, see Appendix B, pp. 204–205 *infra*.

[155] RN, *Life/FN*, p. 242.

[156] RN, *Life/FN*, p. 461, enlarges on this point:

> This resignation to philosofick studys spoyled [Hoskins] the lawyer, but made an accomplisht good companion, especially to one [FN] who delighted in those matters as much as himself. After a long day's work, if his Lordship [FN] could get Sir John

evidence that Hoskins had a keen interest in music, he could have been an intermediary between Hooke and North.[157] But, like the case for John North, that for Hooke is circumstantial. Hence, the reader may form his or her own conclusions from evidence scattered throughout this study.

> Hoskins to a French hous, for a petit souper, but ample feast of discours, he was happy, which I can the better testifye, having often bin one in a company. There was no corner of the univers, that imagination could make accessible, but they searched it to the quick, and nothing new sprang abroad, or at home, but one or other of them early or late, brought it under examination.

The French house at which they dined was Chatelins in Covent Garden. For details about the cost of dinner, see Pepys, *The Diary*, vol. 9, pp. 115, 172, who also indicates that there was music.

[157] Although Hooke's memoranda indicate that he met almost daily with Hoskins to discuss various matters, only one memorandum relates specifically to the philosophy of music as follows: 'With Sir J. Hoskins, Mr. Hill, Dr. Holder, Mr. Aubery, at Sir Chr. Wrens. I propounded my theory of musicall ear'; see 29 April 1676 in Hooke, *The Diary*, p. 229

2 THE SYSTEM OF THE *ESSAY*

In his *Essay* North enters as a philosopher, and his tract itself is an illustration of a certain way of philosophising. His brother, John, had preferred Plato's way,[1] which also had been the preferred way of Descartes, who studied causes and the art of demonstration, seeking epistemological certainty in mathematics. Warning against mixing 'conjectures into the judgements we make about the truth of things', he had recommended the study of mathematics as a way to learn how to argue demonstratively rather than probably.[2] Nevertheless, Descartes conceded that, where truth has yet to be established, we might make any hypothesis about the matters at hand, with the proviso that all the consequences of our assumption agree with our experience.[3] And he had illustrated this way of philosophising in some of the essays published together with his *Discours de la méthode* (1637).[4]

In the essay on dioptrics, for example, Descartes did not rely on mathematical demonstration, because he did not seek the true nature of light. Instead, he aimed to summarise data and suggest properties that are to be assumed—suppositions—even though the properties are unobservable. For this purpose, he relied on a method that consisted of a number of formal analogies—a cane, wine in a perforated vat, a tennis ball and racket—as 'building blocks for representations of direct, reflected, and refracted light'.[5] Although in this essay Descartes presented his explanations as suppositions to be verified by their experimental consequences, he still believed that suppositions are deducible from certain *a priori* principles; and it was this

---

[1] See RN, *Life/JN*, pp. 127–128, 135. JN produced a critical edition of 'Pythagoreorum fragmenta et Heliodori larissæi Capita Opticorum' for inclusion in Thomas Gale's *Opuscula mythologia ethica et physica, Græce et Latine...*(Cambridge, 1671). JN also 'obstetricated' (RN's term) the first critical edition in Greek and Latin of some of Plato's dialogues, which was published as *Platonis de rebus divinis dialogi selecti* (Cambridge, 1671).

[2] Quoted in Franklin, *The Science of Conjecture*, p. 218.

[3] Voss, 'Scientific and Practical Certainty in Descartes'; see also Gozza, 'A Renaissance Mathematics', p. 164, who describes Descartes, *Excellent Compendium*, as a 'model of a demonstrative musical treatise'.

[4] See Descartes, *Discourse on Method, Optics, Geometry, and Meteorology*.

[5] The tennis ball and racket is Descartes' version of the Aristotelian analogy of the motion of a ball after impact; see Chapter 1, p. 17 *supra*; see also Eastwood, 'Descartes on Refraction'.

procedure that led Newton to understand, and later to reject, hypotheses in the Cartesian sense of unverified deductions from first principles.[6]

North's way of philosophising differs from the procedure of Plato and his followers, for his *Essay* is in the form of a 'put-case'—a legal case argued from an hypothesis or supposition.[7] A general principle is assumed, from which all sorts of consequences can be deduced with the aid of logic and other material assumptions; some, if not all, of the consequences are verified empirically; and it is concluded with probability that the principle is applicable to the case at hand. Hence, the title, *Essay*, is best understood as a trial or trial specimen, that is, a test, proof, experiment or assay. In seeking to extract principles relevant to a particular case, North, as a lawyer, had 'by habit a sort of faculty intuitive, so that from symptomes ordinarily litle observed, he conjected the truth of facts, and was thereby guided in the shortest and most proper means by examination to come at them'.[8]

'Symptomes' are signs of facts, sensible as well as insensible; and in law

> evidentiary matters had to be dealt with, and ... decisions based on matters of fact had to be made. As it was increasingly realized that obtaining the truth about matters of fact was not a simple matter either in theory or practice, rules and practices were developed to deal with witnesses, jurors, testimony, and judicial behavior, all of which were aimed at refining the process of reaching truth. Philosophical theories [of knowledge] ... played a substantial role in this process [and] ... were applied in the legal sphere, first somewhat haltingly by [Matthew] Hale and [Francis] North ... [so that by] the eighteenth century, the English legal community possessed a law of evidence that ... rest[ed] on a sound philosophical basis.[9]

But philosophy itself required rational grounds for the decision of facts and for the evaluation of competing hypotheses.[10] Indeed, as North himself intimates (p. iii), the *Essay* was the result of his desire to test a particular hypothesis by tracing his notions 'as near their principles' and setting them down 'in method'

---

[6] For the word, hypothesis, and for Newton's conception and use of hypotheses, see Hanson, 'Hypotheses fingo'.

[7] For a lawyer as a 'put-case', see OED and the citation from RN; see also RN, *Life/FN*, p. 16, where RN writes that FN used the common room in the Temple Hall 'as well at nights as at noons and found advantage by it, for the conversation there runs much upon law, and putting cases (as they call it) and if the same company met at evenings, as when out of comons, their talk was much of that sort'.

[8] RN, *Life/FN*, p. 57.

[9] Shapiro, *Probability and Certainty*, pp. 192–193. For FN's incomplete essay, 'Of Evidence', which begins with the question, 'what is truth?', see RN, *Life/FN*, pp. 58, 393.

[10] For background, see Franklin, *The Science of Conjecture*, especially pp. 1, 131, 195–227.

so as to discover whether 'they were well deduced' not only one from another but also from experiments. But he adds that his experiments are such 'as I had in memory', that he had not made any himself, and that his 'recollections' were based on 'what I heard from others, or occasionally observed'.[11]

Unfortunately, with one exception, no sources are cited in the *Essay*; and this lack of citation presents a problem for the interpreter. Nevertheless, when possible, I shall attempt to identify some of the sources from hints supplied by North himself or from details of his life supplied by Roger North. As for the body of the *Essay* itself, each of its nine sections ('chapters') varies in length; and the brevity of their treatment reflects the lawyer's commonplace method. In his abridgment of the common law, for example, North adopted this method chiefly for the purposes of information retrieval, not because he had a 'bad memory' but because he 'did not think fitt much to trust it'. But he also adopted the method as a way of writing up his researches, for, according to his biographer,

> he found that in his searches after any point, the very casting his eye upon other matters, tho transiently, refreshed his memory of them, which he accounted a great benefit, and for that reason did not over much critiscize upon method, as some doe who use generall heads, and bring the rest under subdivisions. But he [relied on] ... generall heads [without subdivisions]; and sayd he found that way prompter for researches then the other; and this course of commonplacing he used more or less as long as he lived.[12]

The commonplace method disguises the true structure of the *Essay*, which falls into three main parts. In the first part (section 1, pp. 5–7), North offers a conjecture—a brief physical hypothesis about the external events that cause sound. In the second part (sections 2–5, pp. 7–21 and 'Scheme'), he extends the hypothesis to account for the effects of sensory perception, with reference to the mechanics of musical instruments. And in the third part (sections 6–9, pp. 21–35), he develops the implications of his rationale for the perceptual cognition of music in performance and composition with reference to the meanings the mind makes from sensations. In this chapter, I examine

---

[11] For some who might have provided FN with information, see Appendix B *infra*.

[12] RN, *Life/FN*, p. 16. The full quotation relates chiefly to FN's abridgment of the law, which he wrote in 'a small, but very legible hand', a description that fits FN's synopsis of English law —Coke upon Littleton (US:Dlc) KD 833.G85/1685. FN recommended the commonplace method to RN, who used it in many of his unpublished writings, including some of his writings on the philosophy of music. RN also wrote about the practice in an essay on the study of the law; see RN, *A Discourse*. For historical background to the commonplace method in law and other fields, as well as for the transformation of this practice during the seventeenth century, see Moss, *Printed Commonplace-Books*.

North's 'put-case' in each of the sections that constitute the three parts. Before commencing this examination, a number of points need to be made about terminology.

North uses specific terms more or less consistently for the basic elements of his philosophy: 'elasticity' or connate terms (e.g., 'spring') for the physical property inherent in a fluid or solid by which it tends to restore itself to its original position after deformation; 'pulse' or connate terms (e.g., vibration, tremble) for the motion of restitution of invisible bodies, as well as bodies of ordinary experience—especially musical instruments; 'tone' for the sensory effect of such mechanical motions; and 'note' for the sign of our ideas. The last usage is reminiscent of Hobbes, for whom signs are names used by the mind as tokens or counters in computation (ratiocination),[13] so that 'things may be considered, that is, be brought into account either as internal accidents of our mind ... or as species of external things, not as really existing, but appearing only to exist, or to have a being without us'.[14]

According to Roger North, his brother drew up the *Essay* not with 'forme and exactness' but with 'the sence of a man of buissness, who minds the kernell, and not the shell'.[15] This is one way of saying that the formation of concepts adequate to the modern philosophy of music was a slow process, so that some of its central concepts ('the shell')—for example, 'wave', 'interference', 'frequency', 'phase', 'node', 'harmonics', 'simple harmonic motion', 'tonality'—were defined only after the processes involved were better understood ('the kernell'). It is also a reason why some of the terms used in the *Essay* lack precision that ensures certainty of meaning. In what follows I hope to have made clear the basis of my interpretations of such terms, especially in the case of possible disputed readings.

FIRST PART: THE PHYSICAL HYPOTHESIS
'OF SOUND' (pp. 5–7)

North begins the first part of his *Essay* with an enumeration of nine 'phænomena of sound' (p. 5), as follows:

(1) Sound 'may be produced in the Torricellian vacuity';
(2) 'It causes motion in solid bodies';
(3) 'It is diminished by interposition of solid bodies';

[13] On computation, signs, marks and names, see Hobbes, 'Elements of Philosophy', pp. 1-5, 14–24.

[14] Hobbes, 'Elements of Philosophy', p. 92.

[15] RN, *Life/FN*, p. 249.

(4) 'If the bodies interposed are very thick, its passage is wholly obstructed'; and
(5) 'It seems to come to the ear in strait lines when the object is so scituated that it cannot come in a straight line to the ear.'
(6) 'When the air is not in motion its extent is sphærical';
(7) 'When there is a wind, the sphære is enlarged on that part to which the wind blows, and diminished on the contrary part';
(8) 'That it arrives not to the ear in an instant but considerably slower than [light to the organ of] sight';
(9) 'That it comes as quick against the wind as with it, though not so loud nor so far'.

Some of the same phenomena had been identified by Hooke, when, on 25 February 1663, he presented fellows of the Royal Society with a scheme for inquiring into the natural history of the air, with the aim of seeking to answer, among other things, the following questions:

> Whether it [air] be the medium, that conveys sound? By what means sound is conveyed, and how fast? Whether a wind will refract a sound? that is, make the sound seem in another place than really it is. How the sound diminishes according to several distances from the body, that causeth it? How much a fair wind helps the progress of sound, and how much a contrary wind hinders it; and by what means?[16]

About 1666 he revised his queries as follows:

> By what means the Air becomes the Medium of Sound, whether the more dense or rarify'd, moist or dry, transparent or opacous, be fittest for that Conveyance?
> With what Velocity Sound is conveyed through the Air, whether it may not be conveyed swifter by Strings well stretcht, or solid Bodies?
> Whether it proceeds by strait or curve Lines?
> Whether it be not alter'd by Winds, *&c.*?[17]

North's list is neither a set of queries, like Hooke's, nor an assertion of facts, for the truth of his nine phenomena had yet to be demonstrated experimentally. North, therefore, begins his list with a conditional statement: sound 'may be produced in a Torricellian vacuity'.

As previously indicated, the term, 'Torricellian vacuity', was first used after the invention of the barometer to denote the six-inch space at the top of

---

[16] Hooke, 'The Life and Work', vol. 6, p. 114.

[17] Hooke, *The Posthumous Works*, p. 33; for the date, see Kassler and Oldroyd, 'Robert Hooke's Trinity College "Musick Scripts"', p. 582.

the barometric tube.[18] Subsequently, however, the same term was generalised to denote any empty space in experiments with air pumps. Since North uses the term, 'Torricellian vacuity', in this latter sense and since he was a proponent of Boyle's hydrostatic doctrine,[19] it is highly probable that Phenomena 1–5 derive from a chapter entitled 'Experiment XXVII' in Boyle's *New Experiments, physico-mechanicall, touching the Spring of the Air.*[20] There, Boyle took as his point of departure the vacuum experiments on sound, performed from the early part of the seventeenth century by continental philosophers and from which some concluded that not the air but a more subtle body was the medium of sound.[21] In order to 'invalidate such a consequence' from the vacuum experiments, two trials were made in Boyle's laboratory to determine 'what the air doth in conveying of sounds'.

In the first experiment, a watch was suspended by stout twine ('packthread') in the cavity of a receiver or glass vessel, the receiver was closed with melted 'plaister' and the air was left in the receiver. Upon listening near the sides of the receiver, Boyle plainly heard the sound of the watch (made by the motion of its balance). But when he listened near the cover of the receiver, he heard a 'manifest' disparity of sound, which he attributed to the texture of the glass, the cover and the cement through which the sound was propagated. Then, the pump was employed to remove air from the receiver; and as this occurred, 'it seemed, that from time to time, the sound grew fainter and fainter, so that when the receiver was emptied', no sound could be heard from within it.[22]

---

[18] See Chapter 1, p. 19 *supra.*

[19] See Chapter 1, pp. 26–27 *supra.*

[20] For 'Experiment XXVII', see Boyle, 'New Experiments...' [2d edn. 1662], *The Works*, vol. 1, pp. 62–64.

[21] Boyle's only identified source is Kircher, *Musurgia universalis*, vol. 1, pp. 11–13, being the same source used by Charleton, *Physiologia*, p. 229–230, who concluded about the experiment, p. 231:

> whoso[ever] well considers these things, cannot doubt, but that it is much easier to Nature to admit the trajection of Sound produced in the Disseminate Vacuity, through the pores of the Glass-head, than [to admit] the transmission of an External Sound into a close Chamber, through a Glass window....

Charleton, of course, supposed that the medium of sound was 'the subtler or more ætherial part of aer'; see Chapter 1, p. 5 *supra.*

[22] See Lindsay, 'The Story of Acoustics', p. 635, who states that a more careful examination of the phenomena shows that Boyle's conclusion was erroneous, because

> the observed decrease in the intensity of the sound is due not so much to the failure of the low pressure air to transmit sound as to the increasing difficulty of getting the sound out of the bell (or other sound source) into the air and then out again from the air to the glass container. The so-called impedance mismatch between source and surrounding fluid medium becomes greater as the density of the medium increases.

A second experiment was then tried to discover whether or not the absence of air 'hindered us from hearing' the sound. Instead of the watch, Boyle substituted a small bell supported in the middle of the receiver's cavity by a bent stick that 'pressed with its two ends against the opposite parts of the inside of the vessel'. When the receiver was closed up, the bell sounded 'more dead' than it did in the open air; and when the receiver was emptied of air, there appeared to be no 'considerable change' in the loudness of the sound. Hence, Boyle supposed that

> though the air be the principal medium of sound, yet either a more subtle matter may be also a medium of it, or else an ambient body, that contains but very few particles of air, in comparison of those it is easily capable of, is sufficient for that purpose.

And he also considered another possibility, namely, whether

> there might not in the deserted part of the tube remain air enough to produce a sound; since the tubes for the experiment *de vacuo* (not to mention the usual thinness of the glass) being seldom made greater than is requisite, a little air might bear a not inconsiderable proportion to the deserted space: and that also, in the experiment *de vacuo*, as it is wont to be made, there is generally some little air, that gets in from without, or at least store of bubbles, that arise from the body of the quicksilver, or other liquor itself, observations heedfully made have frequently informed us....

In short, the vacuum experiments were inconclusive in determining whether gross air particles or more subtle ones were the medium of sound.

In describing these experiments, Boyle twice referred to the direction of the path traversed by sound. Initially, he observed conditionally that sound 'seemed to come directly in a streight line' from the sounding source to the ear; but, afterwards, he noted unconditionally that 'in a vessel so well closed as our receiver', it was remarkable that such a weak ticking sound from

> the balance of the watch, should propagate a motion to the air in a physically streight line, notwithstanding the interposition of so close a body as glass, especially glass of such thickness as that of our receiver; since by this it seems the air imprisoned in the glass must, by the motion of the balance, be made to beat against the concave part of the receiver, strongly enough to make its convex part beat upon the contiguous air, and so propagate the motion to the listener's ears.

Up to this point, Boyle's observations relate to North's Phenomena 1, 3, 4 and 5. At the close of his chapter, however, he suggested an experiment, but did not perform it, that relates to North's Phenomenon 2, namely,

> whether or no the wirestring of an instrument shut up into our receiver would, when the ambient air was sucked out, at all tremble, if in another instrument held close to it, but without the receiver, a string tuned ... to an unison with it, were briskly touched, and set a vibrating.

The sources for North's remaining four phenomena are more difficult to determine. There are, however, two sources that treat the 'sphære' of sound (Phenomena 6 and 7), its speed (Phenomenon 8) and its propagation with or against the wind (Phenomenon 9). One source is Charleton's *Physiologia*; the other is *Saggi di natvrali esperienze fatte nell' Accademia del Cimento*.[23] The first book has been described as the best presentation of Gassendi's philosophy in England;[24] but the experiments on sound in the second book not only took Gassendi's writings as a point of departure, but also retried Boyle's actual and mooted vacuum experiments on sound. Both books, for example, treat the distance ('sphære', space) that sound travels; and both books report on gun-blast-timing experiments to test the speed of sound. Although there is evidence that North had studied a number of languages, including Italian,[25] it seems probable that Charleton was his principal source.

For example, Charleton, but not the academicians, utilised the term 'sphære'. And Charleton, but not the academicians, referred to Mersenne's computation that although the sphere of light is always equal when a wind blows, the sphere of sound is not equal, for its diameter, when 'heard against the wind, is by almost a third part less than the diameter of the sphere of the same sound, assisted by a favourable ... wind'.[26] This report helps to interpret

---

[23] The 1667 edition of the *Saggi*, published at Florence, was reviewed in the *Philosophical Transactions of the Royal Society* (1668) [No. 33, 17 March], 3: 640. For a translation of the *Saggi*, see Middleton, *The Experimenters*, pp. 83–251, which includes variations on Boyle's vacuum experiments, the results of which were equally inconclusive. When RN began writing on the phenomena of hearing, he made a number of references to the experiments of the academicians, though these references could have been to the 1684 English translation rather than to the Italian original; see (in chronological order) *NP 2*, p. 87; *NP 6*, p. 22; ff. 20v, 21 in 'The Theory of Sound...' (GB:Lbl) 32534: ff. 1–82v; and NP *3*, pp. 61, 74.

[24] See Robert Kargon in Charleton, *Physiologia*, p. xiv. Charleton's many neologisms make the treatise a problem for interpreters and may account for the lack of detailed scholarly attention to his chapter on 'The Nature of a Sound', *Physiologia*, pp. 208–232.

[25] For languages (but not for law French), see RN, *Life/FN*, pp. 14, 45, and 97.

[26] Charleton, *Physiologia*, p. 212; see also *passim* for a number of reports on the speed of sound, including the consideration of how far one must stand from a reflecting body in order to hear an echo. These reports seem to rely on Mersenne; but Lenihan, 'Mersenne and Gassendi', provided evidence that the first attempt to measure the velocity of sound in air was made by Mersenne and that Gassendi knew of Mersenne's work and referred to it in his writings but did not make any experiments himself. For Gassendi's interest to dispel the Aristotelian doctrine that the speed of sound depended on the strength of its source, see Hunt, *Origins in Acoustics*, pp. 95–100.

North's Phenomena 6 and 7. But Charleton also reported on Mersenne's blast-timing experiments to test whether or not sounds are propagated equally swiftly as well as in equal time, regardless of whether the wind is gentle or adverse.[27] Like other philosophers of the time, Mersenne seemed to believe that the velocity of sound in air was independent of such environmental conditions as temperature, humidity and wind velocity.[28] Hence, Charleton, as well as the academicians, could assert that no wind can accelerate or retard sound (North's Phenomenon 9). Even though the accuracy of the tests was jeopardised by a number of difficulties, the experiments would have led to the conclusion that sound arrives more slowly to the ear than light to the eye (North's Phenomenon 8), since the blast-timing method involved counting the oscillations of a pendulum, or the beat of one's own pulse, from the time one observes the flash produced from a gun until the sound arrives at the ear.[29]

Whatever the sources of North's nine phenomena of sound, the main point of his enumeration is that these phenomena, along with the 'visible actions' that produce sound in bodies of ordinary experience, led him to entertain a kinematic variant of the continuum hypothesis. According to him, sound is caused when an impact or driving force shocks a deformable material medium, thereby producing an internal motion ('tremble') that is propagated through the medium to the ear.[30] To this intervening medium, North gives the name 'air' which he conceives as a mixture of minute, invisible bodies of three different sizes: small, middle and gross. The small particles, which fill the interstices of *all* bodies, are 'subtil, ethereal' and incompressible; these are the medium of light. The middle-sized particles are also perfectly hard or impenetrable; and these are the medium of sound. But the gross particles, unlike the small and middle ones, are 'elastical'—air *simpliciter*.

When the whole mixture called 'air' is violently struck, its inelastic particles pass through solid bodies, whereas its gross particles, being elastic, are resisted by solid bodies. Hence, when any of the gross particles are displaced, the inelastic particles enter the vacant space; but when the gross particles, by their elasticity, restore themselves to their former place, the inelastic particles are extruded with violence as from a centre of that place. When the extrusion involves the middle-sized particles, sound results. But North (p. 6) also asserts that solid bodies, when trembling, may produce sound without the presence of gross air particles. This statement has three

---

27 Charleton, *Physiologia*, pp. 221–222. The academicians also conducted such experiments but made no mention of Mersenne.

28 See Lindsay, 'The Story of Acoustics', p. 635, and Hunt, *Origins in Acoustics*, pp. 101–104.

29 For the difficulties, see Hunt, *Origins in Acoustics*, p. 99.

30 By cause, FN means efficient cause (here, elastic impact); see Hobbes, 'Elements of Philosophy', p. 124: 'in all action the beginning and cause are taken for the same thing.'

implications. First, solid bodies are elastic and function in the same way as gross (elastic) air particles: when, upon impact, they are displaced from their equilibrium position, they restore themselves to their original position, so that sound may be produced in a glass receiver (a solid body) from which gross air particles have been evacuated. Second, solid bodies must have vacant spaces or pores through which not only the middle but also the small particles may enter. And third, because elasticity is due to the impacts of inelastic particles that form the small and middle particles of the air's mixture, sound results when *any* elastic body—fluid or solid—is shocked so as to drive out the inelastic bodies as from a centre. In short, in elastic impact, bodies separate after colliding; and this is North's criterion of elasticity.[31]

Since North's kinematic variant of the continuum hypothesis incorporates an explanation of elasticity, we may infer that he understands mechanics as a study of motion against a resistance, whereby motions, when opposed, will order themselves so as to balance or equalise the distribution of force.[32] A pendulum, for example, which is set swinging by the application of a driving force acts against gravity (accelerative force). The applied force displaces the pendulum from its mean position of equilibrium, so that it moves to and fro through the centre of oscillation until it is brought to rest by friction or air resistance (damping force). To keep the pendulum swinging, there must be a constant application of the driving force, which, if applied at the wrong time, might stop the pendulum rather than keep it going. But if a push, for example, is given in the swinging direction each time the pendulum bob reaches its maximum deflection, there will be an energy build-up—resonance—that is directly evident from the increasing amplitude of the deflection of the swing.

In the 1650s Hobbes had modelled his philosophy on the insight that bodies acting under the force of elasticity were similar to pendula.[33] About ten years later, in 1665, when Hooke began to develop Hobbes's philosophy, he utilised stretched strings as models.[34] North's hypothesis, therefore, may be described as either Hobbesian or Hookean. But it cannot be described as Cartesian, since, on the one hand, Descartes' propagated state is an instantaneous tendency to motion, not a real motion that takes time; and, on the other hand, the Cartesian 'third matter', unlike North's, is inert, not elastic: it

---

[31] In Newtonian (classical) mechanics, the criterion of perfectly elastic collision is that bodies always separate with a relative speed equal to their relative speed of approach.

[32] This kind of mechanics was indebted to Galileo's new science of resistance; see Timoschenko, *History of the Strength of Materials*, pp. 11–15, and Kassler, *Inner Music*, pp. 79–85.

[33] See Chapter 1, pp. 10–11 *supra*.

[34] See Kassler, *Inner Music*, pp. 139–143.

undergoes only random motions that are confined to a spherical space by the impact of the surrounding æther.

Although North merely hints that æther is one of the components in the mixture called 'air', in the 'hot' dispute about the barometer, previously mentioned, he claims that the mixture also includes vapours.[35] And in some undated memoranda on different subjects in natural philosophy, he treats the atmospheric mixture as part air, part water and part nitre, the last of which he describes as 'pabulum for flame and breath'.[36] In the same memoranda, when considering the cause of 'springs' or 'elater', North conjectures, but does not definitely conclude, that the elasticity of air, as well as bodies like steel and glass, may be caused by the rapid motion of the æther, for he supposes that this alone permeates ordinary bodies. But he is definite that, when an impact ('violent start') divides the elastic air particles, they rebound—that is, restore themselves to their equilibrium position. Sound therefore is produced 'by restitution, following compression', for it is this motion that 'affects the sphear of sound successively'.

Note that, in the memoranda, North posits a nitrous component that supports both 'flame and breath', so that he seems to have been aware of Hooke's positing of a nitrous component in the air as an essential agent in combustion and respiration. In a treatise published in 1665, for example, Hooke had argued that air was the dissolvent of all combustibles, that dissolution was made 'by a substance inherent, and mixt with the Air', and that the inherent substance was 'like, if not the very same, with that which is fixt in *Salt-peter*'.[37] He also provided an explanation of elasticity as motion against a resistance, for he argued that springy bodies allow the entry of a limited amount of æther into their structure, whereby the structure is held together by a balance between the (vibratory) endeavour of its particles to recede and the opposing ('heterogeneous') compressing vibrations of the ambient medium.[38]

---

[35] See Chapter 1, p. 27 *supra*.

[36] For FN's memoranda, see RN, *Life/FN*, pp. 259–261: 'Of salts and spirits', 'Of springiness or elasticity', 'Hints of inquirys touching the air', and 'Hints relating to the baroscope'. In these mention is made of the Torricellian experiment, 'windgunns' and the air pump ('pneumatick engine'). See also 'Hints for inquirys touching the air', 'More hints. Relating to the baroscope. &c.', (GB: Lbl) Add MS 32510: ff. 265–265v.

[37] Hooke, *Micrographia*, p. 103. See Lysaght, 'Hooke's Theory of Combustion'; see also McKie, 'Some Early Work on Combustion', pp. 161–162.

[38] See Hooke, *Micrographia*, pp. 10–32, especially p. 25, where he presents his theory of congruity and incongruity—in brief, that particles of bulky bodies move congruously with respect to each other but incongruously with respect to the ambient medium. For details of this theory, see Kassler, *Inner Music*, pp. 141–143.

Hooke then applied this explanation to the behaviour of 'glass drops'.[39] These objects, which have the form of a tear-drop tapering to a fine tail, excited much attention in England, for the head withstands hammering on an anvil, while the tail, broken with finger pressure, causes the whole to explode into powder. Although a number of philosophers attempted to account for this phenomenon, it was Hooke who first grasped in principle the modern interpretation.[40] As part of his explanation, he described the core of the drop in tension with an outer shell in compression, which he likened to an arch. He also described the drop's fracture pattern and demonstrated the concept of the starting flaw in the tensile interior of the drop. And he indicated that 'the sudden flying asunder of the parts as soon as this arch is anywhere disordered or broke', proceeds from the springing of the parts, which he compared to a spring.[41]

Here, then, we have an illustration of motion against a resistance. But the drop's fracture introduces an additional element. Because the outer layers are in a state of compression balanced against high tensile stresses in the interior, when the drops are broken, a disintegration front propagates in both directions from the tensile core with a speed similar to that of transverse sound waves in glass. Hooke noticed this additional element, when, in an experiment of breaking the drops in fish glue, he discovered that, like the other drops he had broken, this one 'gave a crack' or brisk noise and also 'gave my hand a pretty brisk impulse'. It is possible, therefore, that North's use of the term 'cracks' (p. 7) as a synonym for impulses stems from his reading of this part of Hooke's 1665 treatise. But this is only by the bye, for North's concern is not with noise but with music, so that he concludes the first part of his *Essay* with the statement (p. 7): 'I shall leave others to apply this hypothesis to the aforementioned phænomena [of sound] (which they may easily do) and proceed to the discourse of musick.'

SECOND PART: MECHANICS OF MUSIC
'OF A TONE' (p. 7)

In outlining his kinematic variant of the continuum hypothesis, North defines sound as a pulse front—a succession or train of short, sharp impulses

---

[39] Hooke, *Micrographia*, pp. 33–44.

[40] See Brodsley, Frank and Steeds, 'Prince Rupert's Drops', pp. 1–2.

[41] Previous to Hooke, Hobbes had explained glass drops by recourse to the analogue of the archer's bow bent, which is similar to an arch; but unlike Hooke, Hobbes did not understand that thermal contraction on cooling is the source of stress; see Kassler *Inner Music*, pp. 198–199.

('cracks'). He makes no claims about the periodicity of this pulse front, perhaps because not all sounds are periodic. But in the second part of his *Essay*, North moves from sound to tone, where the concept of periodicity is central. Like sound, tone is propagated as a 'repetition' of pulses; but for a sound to be musical two things are required. On the one hand, the pulses must be in equal intervals of time; on the other hand, they must be so rapid (closely spaced) as to be insensible and, hence, perceived in sensation as continuous (see also p. 28). The basic element in sensory perception, therefore, is tone, the result of the motion of the medium's elastic particles that swing back and forth with a uniform motion, uninterrupted by jerks, like the bob of a simple pendulum. But North also states that all space is infinitely divisible. Hence, there is a continuum of spaces between pulses—frequencies—extending from those that are insensible to those that are sensible.

North does not use the term 'frequency', but his definition implies that the sensory perception of tone is related to quantity of pulses—the number of complete vibration cycles per unit of time of the stimulus. Even so, there are two ways of conceiving periodic phenomena—phenomena that repeat themselves over and over again. Since all such phenomena possess a characteristic time period, they may be conceived either in terms of the length of time in which one cycle takes place or in terms of frequency, the frequency being the number of times that a complete vibration occurs in an assumed unit of time (for example, vibration cycles per second). It seems to me that North favours the first conception, while having recourse also to the second.[42] But, as indicated, he assumes that vibrations of musical sounds are equal timed—isochronous, for without this assumption, the correlation of pitch and frequency is impossible.

The shortness of the section may be due to the fact that this correlation was already well established in England.[43] Charleton, for example, reported on the association in 1654.[44] Shortly afterwards, Hooke sought to compute directly the frequency of a fly's wings by the sound they made. In 1665 he published a description of his procedure, which began with an attempt to establish the boundary conditions or 'extreme limits' of the wings' vibrations. About these conditions he concluded that

---

[42] According to the OED, frequency, as a term of physics, was coined in the nineteenth century. For an early use of the term, see Charleton, *Physiologia*, p. 222; for RN's first use, see *Cursory Notes of Musicke*, pp. 46, 95, 96, 99, 102, 103, 111, 137, 141, 149. During the seventeenth century, many other terms were used to signify frequency—for example, promptness, sound or pitch of a sound, number or—as in FN's *Essay*—measure of vibrations, etc.

[43] For the continental background, see Dostrovsky, 'Early Vibration Theory', pp. 171–183.

[44] Charleton, *Physiologia*, p. 222, citing Galileo, Mersenne and Gassendi.

> The wing being suppos'd placed in the upmost limits, seems to be put so that the plain of it lies almost *horizontal*, but onely the forepart does dip a little, or is somewhat more deprest; in this position is the wing vibrated or mov'd to the lower limit, being almost arrived at the lower limit, the hinder part of the wing moving somewhat faster then the former, the *Area* of the wing begins to dip behind, and in that posture seems it to be mov'd to the upper limit back again, and thence back again in the first posture, the former part of the *Area* dipping again, as it is moved downwards by means of the quicker motion of the main stem which terminates or edges the forepart of the wing.[45]

Supposing that the buzz 'proceeds from the exceeding swift beating of the air by the small wing', Hooke sought to estimate the frequency of the wings' vibrations between the established limits by comparing the pitch of the buzz with that produced by a 'musical string, tun'd unison to it'. Presumably, although he does not say so, he then attempted to find out the number of vibrations per second or other assumed unit of time by counting the vibrations of the string. Since he would have known that the more rapid the vibrations, the higher the pitch, Hooke could state that the buzz of a bee is shriller, because 'the wing of a Bee is less in proportion to its body, then the other wing to the body of a Fly'. Hence, he speculated that there may also be other insects that afford an even 'more shrill noise' with their wings, so that 'the quickest vibrating *spontaneous* motion is to be found in the wing of some creature' yet to be discovered. This last statement provides a context for North's statement about infinite divisibility, for it implies that frequency is an infinite series or mathematical continuum, so that there is no rate so slow that there might not be a slower, nor so fast that it could not be faster. In the case of Hooke's flying insects, for example, some may have frequencies so quick that their buzz is beyond the range of human audibility.

'Of the Relation of Tones and the Union of Mixt Sounds' (pp. 7–11)

North extends the definition of a tone to show how consonance is formed. But he is particularly concerned to provide a physical justification for the relative perfection of consonances or, as he more consistently writes, 'chords' (see p. 3). This term is not to be understood as a harmonic structure of three or more simultaneous sounds, even though in some cases this meaning is implied. Rather, North uses 'chord' to mean: (1) concordant intervals in general and as applied to (2) a simultaneous set of simple tones present in a complex tone, and to (3) a successive series of complex tones that can be produced from a single

---

[45] Hooke, *Micrographia*, pp. 172–173.

fundamental on stringed instruments (flageolet tones) or on wind instruments (overblown tones).[46] In this section, North uses 'chord' in both its first and second meaning, as will appear.[47]

Before 1677, some philosophers had attempted to provide justifications for intervallic relations. Excluding number mysticism, two main approaches may be identified.[48] One approach, called the '*senario*', used proportions up to the number six to compare lengths of vibrating strings,[49] so that the proportions index the relational structures not of sense perceptions but of strings and other sound sources. Descartes used this approach when he derived consonances from the senary division of the string.[50] But he was inconsistent in his recourse to senary justification, asserting, for example, that 'whether we extract the perfection of Consonances from Division [of the monochord], or from *Numbers*', some intervals are just pleasant or unpleasant to the ear.[51] Since he refused to comment on how 'men shall imagine the sound to be perceived by the Hearing', he resorted to an explanation that relied on the ease with which the listener distinguishes certain arithmetical proportions,[52] stating that

> I can subdivide the line *A B* into 4, or 5, or 6 parts, but no further; because such is the imbecillity of the Ears, as that they cannot distinguish, without so much labour as must drown the pleasure, any more Differences of Sounds.[53]

---

[46] Five phenomena, identified by Green, *The Harmonic Series*, p. 97, could have provided FN with the experience of secondary tones above a fundamental tone: (1) the natural series of flageolet tones; (2) the natural series of overblown tones; (3) the observation of individual partials in a complex tone; (4) the synthesis of tonal mixtures as in compound stops in an organ; and (5) resonance. Some experience would have been direct, since, according to RN, *Life/FN*, pp. 12, 13, 129, 248, FN was a performer on both bass and lyra viol. Of the latter instrument, there were two types. The common type produced maximum resonance of the open strings by means of special tunings, whereas the uncommon type (out of favour by c.1660) produced resonance by a course of sympathetic strings placed within the neck.

[47] According to Helmholtz, *On the Sensations of Tone*, p. 24, 'every musical tone in which harmonic upper partial tones can be distinguished, although produced by a single instrument, may really be considered as in itself a chord', that is, as a harmonic structure of three or more sounds.

[48] For number mysticism, see, e.g., Mace, *Musick's Monument*, pp. 264–268; and Simpson, *The Division-Viol*, pp. 23–24.

[49] For the *senario*, see Shirlaw, *Theory of Harmony*, pp. 37–44, and Cohen, *Quantifying Music*, pp. 3–7.

[50] See Shirlaw, *Theory of Harmony*, pp. 61–62, and Cohen, *Quantifying Music*, p. 163.

[51] Descartes, *Excellent Compendium*, pp. 19–20.

[52] Descartes, *Excellent Compendium*, p. 23.

[53] Descartes, *Excellent Compendium*, p. 9.

We may conclude, therefore, that Descartes' principal justification refers to a 'normal' sense perception that occurs only when there is a proportion or due relation between the properties of the sound source and the limitations of the senses. This is the norm, and it is defined by the ease with which an object and its constituent elements can be perceived. But behind this justification stands the *a priori* principle that ease of sense perception is given only by the perfect arithmetical proportions—the Pythagorean simple numbers 2, 3 and 5.[54]

A different approach was taken by some continental philosophers, who justified consonance physically on the basis of the coincidence of motion of two vibrating strings—the so-called 'coincidence theory' discussed in Chapter 1. In spite of some problems, to be raised later on, this theory is historically important, because it facilitated an empirical explanation of consonance and dissonance in physico-mathematical terms, thereby undermining the traditional rationalist approach. As previously indicated, there were two versions of the theory within the context of continuum hypotheses. One version considered the total number of concurrences between strings without regard to time, whereas a second version placed intervals in an order of consonance based on the time required for consonance to achieve full cycle. Galileo was one of the continental philosophers who developed the second version; and Hobbes presented an interpretation of Galileo's justification to English readers.

It is important to note that Galileo ordered the consonances in the same way that the purely arithmetic Pythagorean scheme orders them, by multiplying the terms of the frequency ratios of the consonant intervals.[55] This gave the following order: octave, fifth, fourth, major sixth, major third, minor third, minor sixth. According to Floris Cohen, this order 'follows inexorably from the coincidence theory'.[56] But North (p. 9) gives the following order: octave, fifth, fourth, major third, minor third, major sixth, minor sixth. Why, then, is his order different from that of Galileo? The answer seems to be that North presents a third version of the coincidence theory, one that ignores number proportions and explains consonance and dissonance in terms of the sensory perception of periodic pulse fronts in the air. As will appear from his 'Scheme' (p. 3), to be examined presently, he finds the relative perfection of

[54] According to Cohen, *Quantifying Music*, p. 165, Descartes' attempt to base the scale on the numbers 2, 3 and 5 was 'new'. But Descartes' approach followed that of the ancient Pythagoreans, who calculated the scale of music employing ratios based only on the prime factors of 2 and 3; see Kassler, *Music, Science, Philosophy*, pp. 31–35.

[55] Mersenne used this, as well as another method, for which see Gruber, 'Mersenne and Evolving Tonal Theory', pp. 45–46.

[56] Cohen, *Quantifying Music*, pp. 94–95.

consonances in the coincidence of vibrations observed in the fundamental and five secondary tones of the harmonic series.[57]

North's 'Scheme' elucidates the related text on the formation of consonance, which begins thus: 'When the pulses of tones are coincident one with the other, there is a union of the sounds.' This is North's brief definition of consonance as the effect of two or more tones that unite, that is to say, co-exist undisturbed in the ear. He follows this definition with three important statements. The first statement (p. 7) concerns elastic bodies—when moved, they 'vibrate by intervals, according to the measure of the spring until the force be spent'. The word 'spring' denotes elasticity—the restoring force that controls the timing or frequency. When the restoring force is proportional to the displacement, an elastic body will swing with a regular type of pendular or simple harmonic motion, one that repeats over and over again in more or less equal intervals of time.[58] The second statement (p. 8) concerns resonance—the reinforcement or prolongation of sound by synchronised vibrations, for North points out that a periodic support of vibrations increases their amplitude and, hence, their energy build-up, until they become 'sensible' to both eye and ear, whereas a lack of such support prevents the production of sound. And the last statement, made somewhat later in the section (p. 10), concerns dissonance, an effect that North attributes to periodic fluctuations of loudness that we now call 'beats'.[59]

From these statements we may conclude two things. First, the ratios that determine intervals are to be found in the natural relations that may be observed to exist in the resonance of elastic bodies such as stretched strings. Second, underlying North's explanation of the reinforcement and destruction of sound is a mechanical principle—that different modes of periodic vibration

---

[57] For the term 'harmonic series', see Chapter 1, p. 2 n.5 *supra*. In 1701 Joseph Sauveur coined the term *sons harmoniques* to denote those secondary tones produced by frequencies that are integral multiples of the fundamental frequency.

[58] Simple harmonic motion is a term of late (perhaps nineteenth century) coinage. In 1543 Nicolaus Copernicus had analysed this form of motion as the product of two combined circular motions; and he noted that it is analogous to suspended bodies like pendula; see Bell, *Christian Huygens*, p. 99. Hence, Copernicus may be one of the sources for Hobbes's conception of 'simple motion'; see 'Elements of Philosophy', p. 320, and Kassler, *Inner Music*, pp. 56–57. On 21 November 1666 Hooke's account of inclining pendulums included a comment on the conditions required for simple harmonic motion: that the 'equality of duration of vibrations of differing arches [i.e., arcs] or lengths depends upon the figure of the curve-line, in which the body is moved; which figure being for a very great part near the same with that of a circle, it follows that, the motion in differing arches of the same circle will be very near of equal duration.' See Birch, *History of the Royal Society*, vol. 2, pp. 126–127.

[59] For his names, see Appendix A, pp. 143–44. Beats are produced by the superposition of two (or more) tones of close, but not identical frequencies. According to Maley, *The Theory of Beats*, p. 20, the 'explicit identification of audible beats as the cause of dissonance is rare', so that he seems to believe that Sauveur was the first to make this identification.

occur simultaneously within the same vibrating body and that these different modes mutually influence each other's period and force themselves into synchronism. This principle implies, in turn, a new definition of tone as a compound or sum of simple tones and, hence (pp. 17–18), as a composition of simple pendular vibrations.

The first part of this new definition appears in the 'Scheme', where a figure in musical notation (*Figure 1.B*) represents the first six notes of the harmonic series arranged as a harmonic structure that is equivalent to what we now call a major chord over a fundamental bass. Adjacent to the notes is a series of Arabic numerals ascending from 1 to 6. These numbers are not to be interpreted as the arithmetical progression of the *senario*, for they represent frequencies—the number of pulses of the higher tones that are synchronous with the lowest tone. For example, the first note, the octave above the fundamental, makes double the number of vibrations in the same time as the fundamental; the second note, or fifth above that octave, makes three times as many vibrations in the same time as the fundamental, and so on. Hence, the numbers 1 to 6 show how many times the upper note exceeds the fundamental in vibrations. North (p. 9) proves that the constituents of this harmonic structure have their 'pulses in such proportionable time' by the sectional division of the total string (1) into 1/2, 1/3, 1/4, 1/5, 1/6 parts. If we interpret his numerical series written both as frequencies and as string lengths, the implied rule is: frequency of vibration is the inverse of string length, so that, for example, halving the string length doubles the frequency.

Since frequencies that produce consonances are too rapid or closely spaced for conscious counting, North's 'Scheme' includes two graphs, Tab. i and ii (*Figure 1.A*). In these symbolic diagrams, time is the horizontal axis and frequency, the vertical axis, the dots on the latter representing the proportional times ('symmetry') of consonances in relation to the fundamental tone, now including the major and minor sixth as inversions ('reverses') of the major and minor third.[60] There are, however, a few points to note about these graphs. First, Tab. i is divided into seven segments by means of a continuous line drawn vertically across the page and in which the second row of dots in the first segment represents the pulses of more than one octave. Second, the dots on the lowest part of each of the segments show the relation to the fundamental tone (1) in the topmost segment, as is seen more clearly in the reduction represented in Tab. ii (the fundamental occurs on every eighth vertical line). Third, in Tab. i, segment 2 downwards, the dots on the upper parts of each segment show each individual consonance, as for example, the periodicity pattern of the fifth, 3 to 2. Fourth, although both graphs represent the pulses

[60] See Helmholtz, *On the Sensations of Tone*, p. 14, for the method of finding the corresponding ratios of the frequency (pitch) numbers, 1–6.

beginning at the same instant, this is merely a graphical convenience, for, according to North (p. 8), when one of the strings commences its vibrations after the other, 'by the second or third pulse' their vibrations would become coincident because of resonance. Fifth, in the more detailed Tab. i, North represents the intermediate 'by-pulses' not by dots only but by dots connected to a small vertical line placed before, after or on both sides of the dot.

Because North explains consonance and dissonance in terms of the sensory perception of synchronous pulses—pulses that strike the ear in the same time, he also has to explain why asynchronous 'by-pulses' cancel the pleasure brought by the union of synchronous pulses. His explanation (pp. 9–10) is that the oppositions balance each other, because the intermediate by-pulses

> are all placed in such manner, in relation to one another, that where any of them distract the pulses of the concording string on the one hand, there are others that ... set them right ... so that it is as if there were no intermediate pulse[s] at all, as to any effect of hindring the coincidence.[61]

North does not account for the 'torment' that might be caused by a lack of synchronisation between the pulses, because his subject is harmonious intervals, not disharmonious noise; and, as he later indicates (p. 32), harmony is designed to please, not to torment the ear.

Nevertheless, his 'Scheme' is historically important in five different ways. First, it shows that, on North's principle of relative perfection of consonances by coincidence of pulses, the fundamental tone is the norm with which the pulses coincide, thereby providing a rationale for a fundamental bass. Second, it justifies a statement in the text (p. 10) that 'all octaves ... are as it were included in any tone',[62] thereby providing a rationale for octave identity. Third, the frequency numbers of *Figure 1.B* support a physical justification of the perfection of the major chord, thereby providing a rationale for chords as harmonic structures of three or more simultaneous sounds. Fourth, all these physical justifications constitute the rationale for the 'systeme'

---

[61] This statement, though cryptic, is a consequence of FN's understanding of mechanics as motion against a resistance. According to Chenette, *Music Theory*, p. 187, the statement implies that a similar balancing would take place in all intervals and leave them with similar indexes of consonance, 'for the average amplitude of any balanced vibrating system is zero'.

[62] Given the context, FN's statement refers to the first tone above the fundamental—the first harmonic (Sauveur) or partial (Helmholtz). According to a number of scholars, e.g., Lenoble, *Mersenne*, p. 484, the earliest discussions of harmonics are to be found in statements about the octave by the Aristotelian author of the *Problemata*, by Mersenne and by Descartes; but see Green, *The Harmonic Series*, pp. 318–326, has provided a well-evidenced argument against this claim.

that North claims to have established (p. iv) and that he outlines in the third part of his *Essay*.

Finally, Tab. i and ii (*Figure 1.A*) may be regarded as precursors of the twentieth-century sonogram—a graphical representation of the distribution of sound energy among different frequencies, especially as a function of time. Of course, North's symbolic diagrams are more laborious to produce than a sonogram; but his spatial representation of temporal phenomena requires pen and paper only, so that about 1717 Roger North began to develop this 'method of punctation',[63] since he believed that it was 'the best I have knowne used to depict the consonancys of tones', for 'it is not onely instructive but pleasant, becaus ideas of things withdrawne from sence, are difficult to forme in one's mind' except 'by the help of such images'.[64] Unfortunately, little attention has been paid to representations of physical aspects of sound before the advent of electronics made this task easier. Perhaps because of this lacuna in research, Carlton Maley supposed that Leonhard Euler was the first to represent sound pulses by dots,[65] although Euler's representation appeared after the death of Roger North in 1734.[66]

'How tones are produced, and of assistances to the sound by instruments' (pp. 11–16)

North turns next to the generation of tone in musical instruments, which he defines as bodies that 'stand upon a spring'—the elastic force that controls the timing or frequency. He provides three examples of springy bodies—bells,[67] strings and air columns in pipes. Even though his focus here and in the next section is on strings and pipes, his brief mention (p. 11) that bells 'vibrate by cross ovals' is intriguing, because, to the best of my knowledge, this vibratory pattern was first demonstrated by Hooke at meetings of the Royal Society on 9,

---

[63] See ff. 30v, 32–35v, 37, 41, 41v, 43, 47 of RN, 'The Theory of Sounds...', (GB:Lbl) Add MS 32534: ff. 1–82v; for the later texts, see also *NP 3*, p. 139.

[64] See f. 30v of RN, 'The Theory of Sounds...', (GB:Lbl) Add MS 32534: ff. 1–82v.

[65] See Maley, *The Theory of Beats*, p. 59, and Euler, *Tentamen*.

[66] For some later 'punctationists', see Smith, *Harmonics*; Chladni, *Die Akustik*; Herschel, 'Sound', *Encyclopædia metropolitana*, vol. 1 (plates), figs. 29–35; Higgins, *The Philosophy of Sound*, p. 72; and Curwen, *A Tract*, pp. 15–22.

[67] FN recognises (p. 16) that without the effect of damping, bells will echo their own tones, i.e., produce a prominent strike tone accompanied by a number of distinctly identifiable secondary tones in a composite 'hum' that continues after the strike tone dies away. For the acoustic spectrum of bells, see Green, *The Harmonic Series*, pp. 288, 292.

23, 30 March 1671.[68] North's concern, however, is to make good his earlier promise (p. 7), namely, to show how elastic impact—the 'action' that causes sound—'is performed by the several instruments of musick'.[69] To fulfil this promise, North treats musical instruments as vibrating systems that perform two kinds of vibration—free and forced.

When a single impact displaces a vibrating system from its equilibrium position, it vibrates with a frequency characteristic of the system; the amplitude decays gradually, depending on the resistance of the medium to the motion and on the inertia (mass) of the system, until the energy supplied by the initial displacement has been expended into the medium. North uses the term 'natural vibrations' (p. 15) to denote the frequency characteristic of a freely vibrating system; and he refers to the effect of this natural frequency as the 'tone', 'natural tone' or 'true tone'—that is, the fundamental frequency. When a driving force acts periodically on a system capable of vibrating, we have forced vibrations—forced, for example, by blowing, bowing, plucking or striking a musical instrument. The resulting vibrations consist of a transient component of frequency given by the natural frequency of the system and a steady component of frequency equal to that of the driving force. The amplitude of the steady vibration is a maximum when the frequency of the driving force is equal to—synchronised or keeps time with—the natural frequency of the system without the effect of damping. This synchronised condition is called 'resonance', and the vibrating system is said to resonate with the periodic support—the driving force.

For the generation of tone in musical instruments, however, there are two requisites. First, there must be an acoustic vibrator—for example, gut or wire strings, metal or cane tongues, a player's lips, thin films. When excited, these elastic bodies are the chief source of timing control, for change of frequency is synonymous with change of pitch. Second, because the sound produced by acoustic vibrators is scarcely audible, they must be fitted to resonators or acoustic levers in the form of wooden cases to reinforce the vibrations. By means of this 'continuity' (p. 16), a musical instrument becomes a coupled system of acoustic vibrators and resonators. North recognises (p. 15) that because the sound produced by acoustic vibrators is weak, they must be coupled to resonators so as to transmit their vibrations to the medium. He instances the continuity of coupled systems, first, with reference to the soundpost of a violin, which is a dowel of soft wood inserted upright in the body and serves to transmit synchronous vibrations from the belly to the back of the instrument; and, second, with reference to hearing by bone conduction,

---

[68] Using a glass bell containing flour, Hooke showed the vibratory patterns of a bell by striking or by bowing with a fiddle stick the sides of the glass; see Kassler, *Inner Music*, pp. 143–147.

[69] See this Chapter, p. 45 *supra*.

where a stick similarly serves to transmit synchronous vibrations from a musical instrument to the jaw ('teeth') and, thence, to the auditory nerve of a person who has stopped his ears.[70]

Not all sound vessels are resonators. It is worth noticing, therefore, that North begins his examination of the initiation of vibrations with reference to aerophones, where air is the resonator. Today, we divide aerophones into four basic categories: (1) free aerophones, in which the vibrating air is not contained in the body of the instrument, as, for example, in certain organ reed stops; (2) flutes, where air impinges upon the sharp edge of an aperture; (3) reed pipes, in which air is set in motion by a vibrating reed; and (4) lip-vibrated instruments, such as trumpets and horns. Before 1677, however, little was understood about the operation of aerophones and, in particular, the nature of a wave in the air and the way in which it fits into a pipe. A number of continental philosophers attempted to solve this problem without success.[71] And by the 1650s some English philosophers also were considering the problem. In 1654, for example, Charleton supposed that in wind instruments, 'the stroke of the aer against the sides of the Concave causeth Sound' and that the 'sides of the Concave' are 'very often impulst and repulst' by 'quick reverberations' of the 'inflated Aer'.[72] Not long afterwards, Hobbes concluded that,

> if the breath, when it is blown into a pipe, do only rake its concave superficies, or fall upon it with a very sharp angle of incidence, the sound will not be clear, but hoarse. But if the angle be great enough, the percussion, which is made against one of the hollow sides, will be reverberated to the opposite side; and so successive repercussions will be made from side to side, till at last the whole concave superficies of the pipe be put into motion; which motion will be reciprocated as it is in collision; and from this reciprocation being propagated to the organ [the

---

[70] In performance, especially singing and string playing, the inner ear receives two types of vibrations: one produced by external air conduction, the other by internal bone conduction. Hearing by bone conduction is a combined effect of compressions of the entire skull and of the auditory meatus, together with shaking movements of the skull with a complicated frequency pattern; for details, see Békésy, *Sensory Inhibition*, pp. 120, 193, 195–197, and 'Improved Musical Dynamics', pp. 149, 153–156. In the 1640s the physician, John Bulwer, noticed how the deaf enjoyed music through their teeth; see Mullett, '"An Arte to Make the Dumbe to Speake, the Deafe to Heare"', pp. 132. In 1668 William Holder, described how his deaf pupil had 'some perception of any very vehement sound' when he held one end of a lute string in his teeth; see Holder, 'Account of an Experiment, concerning Deafness', p. 666. And in 1676 Thomas Mace, who had a '*Great Defect in Hearing*', dispensed with the '*Help of My Teeth*' by redesigning the lute so that, by enlarging the sound vessel and by adding more strings, its sound would be augmented through resonance; see Mace, *Musick's Monument*, pp. 203–206. For a later instance, see Kassler, *The Science of Music*, vol. 2, pp. 786–787.

[71] For the investigations of Mersenne, Galileo and Huygens, see Dostrovsky, 'Early Vibration Theory', pp. 190–193.

[72] Charleton, *Physiologia*, p. 220.

> auditory system], from the reaction of the organ will arise a clear sound, such as is made by collision, or by breaking asunder of hard bodies.[73]

North improves on these as well as on previous attempts to understand the acoustics of pipes. He recognises that the operation of aerophones is based on synchronous vibrations;[74] and he distinguishes two different methods by which such vibrations are excited. In one method (pp. 13–14), a stream of air ('thin film'), issuing from a slit, flows out of the mouth past a sharp-edged lip ('solid edge') and produces a vortex motion ('eddy') of the outer air. Although North is not strictly accurate in describing this method,[75] he grasps that the edge-tone mechanism pertains to flue pipes in organs [*Figure 2.1*]. However,

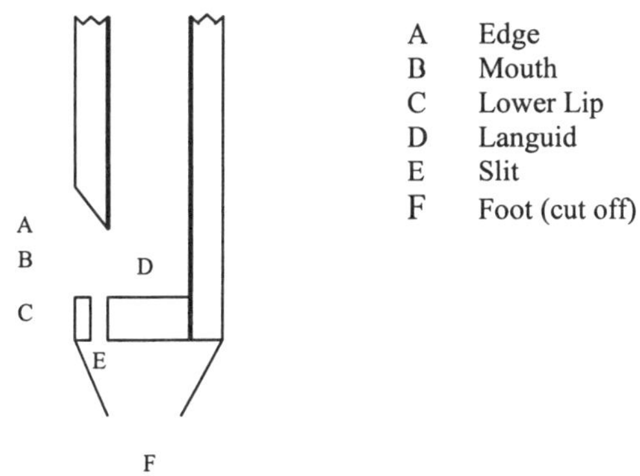

Flue pipe

Figure 2.1

---

[73] Hobbes, 'Elements of Philosophy', pp. 492–493.

[74] In the linear theory of pipes, exemplified by FN, the air in the pipe is conceived as a conservative system, and amplitude is independent of frequency.

[75] See Chapter 3, pp. 113–115 *infra*.

he seems unaware (p. 14) that it applies also to whistles and flageolets or recorders. In the other method (pp. 14–15), a stream of air is periodically interrupted by means of a resilient tongue known as a reed ('quill', 'spring'); and this method pertains to reed pipes in organs, as well as to aerophones with resilient tongues or where the lips of the player perform the office of an acoustic vibrator.

In addition to methods of excitation, the acoustic spectrum of an organ pipe—North's chief focus—is influenced by such variables as length, diameter and shape of the pipe; configuration of the mouth; thickness of the walls and density of the material; size of the foot hole; and magnitude of the wind pressure. North rules out (p. 12) the material from which pipes are made as a determinant of tone.[76] Instead, he identifies three variables: (1) the pipe's dimensions, specifically cross-section ('diameter', 'bore') and length ('bigness'); (2) velocity of efflux due to the strength of the blast; and (3) wind ('air') pressure. Unfortunately, he uses the term 'gage' to refer to one or more of three linear measures, so that the relation of the first determinant—the ratio of cross-section to length—although hinted (p. 19), is not explicit. But North provides qualitative statements about another relation (pp. 13–15), when he indicates that a change to the dimensions of the included air will alter the tone. Ten years later, the correlation between wavelength and pipe length would be stated explicitly by Newton.[77]

Today, we know that frequency depends on two factors: (1) the mass (inertia) of the vibrator, and (2) its elasticity. For a given mass, the greater the elasticity, the higher the frequency will be; whereas for a given elasticity, the larger the mass, the lower the frequency will be. Although most vibrating systems are more complex than this mass-spring combination, the general statement regarding frequency still holds. In 1677, however, definitions of mass and elasticity were unavailable, until Newton defined mass in 1687 and Hooke defined elasticity in 1678.[78] Hence, in a brief statement about the elasticity ('spring') of the air, North observes (p. 12) that the pressures ('compressing force') and expansions (restoring force) are proportionable. In 1661 Richard Towneley noticed this reciprocal correlation between density ('weight') and pressure; in 1660 and 1662 Boyle published an incomplete, then

---

[76] This statement seems to be true, although there has been widespread debate about the effect of the material of pipes; see, e.g., Backus, *The Acoustical Foundations of Music*, p. 211.

[77] See Chapter 3, p. 100 *infra*.

[78] Newton, *Mathematical Principles*, Definition I, p. 1, defined mass as density multiplied by volume. According to Alexander (ed.), *The Leibniz-Clarke Correspondence*, pp. li–liii, Newton never provided a clear definition of inertial mass.

a complete statement of it; about the same time it was verified by Hooke, who published his verification in 1665.[79]

North's treatment of aerophones seems to be guided by his understanding of organ pipes, about which he will have more to say in his next section. Although it now is difficult to discover how he came by this knowledge, his biographer provides some clues—for example, that he was 'adept' in mechanics,[80] spending 'whole afternoons in going about London, onely to see how divers trades worked'.[81] It is probable that these peregrinations included visits to organ builders, as is hinted in the *Essay* itself (e.g., pp. 9, 15). Since North had strong views about religions that were not tolerated by the state, we might rule out a builder who was supported by the Roman Catholic faction at court.[82] A possible candidate, therefore, might be the Protestant, Bernard Smith, who, by 1671, was organ builder to Charles II, and who is known to have had contact with the North family.[83]

In 1675 North, accompanied by his brother, Roger, rode the western circuit as a judge, when he made notes on the Exeter Cathedral organ.[84] These notes, transcribed by Roger North, include a description of the organ's largest pipes—a double diapason (double octave) with speaking length, 20 feet 6 inches; 'nose', 4 feet; circumference, 3 feet 11 inches; and diameter, 1 foot 3 inches.[85] Being slow to speak, the double diapasons were assisted by wood pipes placed behind and tuned an octave higher, presumably, the feature described in the *Essay*, where Francis North writes (p. 20):

> An organ-pipe of a very deep base will not speak suddenly when it is alone; but if an octave be in play at the same time, it will answer the

---

79 See Boyle, 'New Experiments...' [2d edn. 1662], *The Works*, vol. 1, pp. 156–160; and Hooke, *Micrographia*, pp. 222–225. For Townley's contributions and for Hooke's verification, see Webster, 'The Discovery of Boyle's Law', pp. 470–472, 486–495.

80 RN, *Notes of Me*, pp. 100, 110.

81 RN, *Life/FN*, p. 18.

82 According to Grassby, *The English Gentleman in Trade*, p. 201, FN arranged a marriage for his eldest son while still an infant to prevent him from marrying a Catholic in his maturity. Hence, we may rule out the organ builder, Renatus Harris, who was supported by the Catholic faction.

83 Bernard Smith, a German active in Holland from the 1650s, settled in England by 1667. For his work at court, see Ashbee and Lasocki, *A Biographical Dictionary of English Court Musicians*, vol. 2, pp. 1019–1021.

84 The organ was built in 1665 by John Loosemore, who may have been related to the two organists of the same surname employed in the 1660s by FN's grandfather.

85 For FN's notes, see (GB:Lbl) Add MS 32518: f. 102; for RN's transcriptions, see *Life/FN*, p. 63; see also Wilson, *Roger North*, p. 39, for RN's earlier draft of this passage, which differs somewhat in its detail.

> quickest touch, because the tone of the lesser pipe being in chord causes a motion of the air included in the greater pipe, and the vibrations of the included air being coincident with those of the tone of the pipe, promote its speaking by putting the pulses at the mouth into a regular motion, or at le[a]st assisting thereto.

We may infer, therefore, that the great Exeter pipes were open flue pipes that overblow the octave (not, as closed pipes, the twelfth) and, hence, must be made twice as long as the desired speaking length.

'Of the Varying and Breaking of Tones' (pp. 17–21)

North's first mention of overblowing (p. 15) occurs with reference to the process whereby a performer on the trumpet, by adjusting wind pressure and lip tension, obtains a sequence of overblown tones from a single length of tubing. In terms of the mechanics of vibration, this process brings about the conditions for higher modes of vibration in the air column, where the column 'breaks' into separate vibrating segments, thereby varying the pitch. Before the invention of valves, musicians designated the sequence of overblown tones as 'trumpet notes', just as North does (p. 19). For musicians, it was easy to recognise the pitches of a trumpet's natural series; but it was not easy to understand the way an air column vibrates. Hence, as Burdette Green has observed, an accurate explanation of overblown tones was not formulated until the more easily grasped principles of the vibrating string were applied to the natural, valveless trumpet.[86]

Note, therefore, that North begins the subject of segmental vibration ('breaking') with the 'perfect' string that is necessary to produce higher 'chords' (p. 18)—here with the meaning flageolet tones.[87] Prior to 1677 treatments of strings were practical, the aim being to assist lute and viol players in the purchase of new strings by providing a rule of thumb for distinguishing between so-called 'true' and 'false' strings.[88] North's treatment is completely different, for he seeks to provide a rationale from mechanics by referring (p.

---

[86] Green, *The Harmonic Series*, p. 210.

[87] I.e., a successive series of complex tones that can be produced from a single fundamental on a stringed instrument. FN states (p. 17) that flageolet tones are 'not so gratefull to the ear, nor much in use', a statement that is true for his day, since they were not extensively used in musical practice until after 1738; see Green, *The Harmonic Series*, p. 27.

[88] For rules of thumb, see, e.g., Mersenne, *Harmonie universelle*, pp. 79, 188–189, and Mace, *Musick's Monument*, p. 67. From 1677 treatments of strings would become more scientific as it was gradually understood that a string tends to distort the natural frequencies of its harmonics and that the distortion is brought about by the stiffness factor of the stringing material (e.g., brass, gut).

17) to 'the laws of a pendulum'.[89] An attempt to formulate these laws began with Galileo, who made several claims about the pendulum: that its oscillation is equal timed or nearly so; that the period of oscillation is determined by the length of the pendulum; and that this period is independent of the amplitude of oscillation and of the weight of the bob. In 1638 he improved the second claim, which is his law of length, according to which the period depends on the square root of the length of the pendulum—if the length is constant, the period is also constant, that is, equal timed. And he gave an experimental justification of his law in a discussion of problems in music.[90] These claims will help to understand North's procedure, which begins with certain kinds of physical pendula as analogues for true and false strings.

For the true or perfect string—one that produces a tone, North takes a simple string pendulum hung with a bob [*Figure 2.2*]. When this pendulum is removed from its equilibrium position and left free, it will swing to and fro

Simple string pendulum

Figure 2.2

[89] There are four laws: period is independent of amplitude, independent of mass, varies as square root of length, and oscillations are isochronic—equal timed—for a given length; see Ariotti, 'Aspects of the Conception and Development of the Pendulum', also for the problem of the isochronal curve.

[90] Galilei, *Two New Sciences*, p. 107; see also Ariotti, 'Aspects of the Conception and Development of the Pendulum', pp. 350–366.

through the centre of oscillation with a continuous and regular motion. The period of oscillation of such a pendulum is constant; that is, successive oscillations of the same pendulum are equal timed or nearly so. For the false string—one that produces a 'horrible jarr', North takes a simple string pendulum with an added weight fastened near the top of the string, so that two independent vibrations will exist in this system [*Figure 2.3*]. Galileo was unable to account for the behaviour of a non-rigid compound pendulum with two small masses suspended along one thread.[91] But North accounts for it by supposing that each bob will persist in its period at the cost of obstructing one another's oscillations, so that the motions of the individual vibrating segments will be asynchronous—out of time, one with another.[92]

Non-rigid compound pendulum

Figure 2.3

Having set up these analogues, North continues (pp 18–19):

> as a string by such small accidents is brought to vibrations of several parts of the string that hold no proportion one to another, much more may it be imagined to have vibrations of uniform parts, that shall be synchronous, and produce one single sound; as if the parts that vibrate severally are half parts, seconds or thirds of the string. And how this may be caused, I will enquire.

[91] See Galilei, *Dialogue concerning the Two Chief World Systems*, pp. 230–231.

[92] For the problems and for the solution, which required a rigid connection between the two masses, see Ariotti, 'Aspects of the Conception and Development of the Pendulum', pp. 355–357, and Bell, *Christian Huygens*, pp. 150–153.

To convince the reader that tone is a linear superposition of fundamental and secondary tones, he needs to show that a true string has a composition of synchronous vibrations, each segment of which moves as a simple pendulum. For this purpose, North introduces a long string pendulum held fast at the top and hung at the bottom with a heavy bob so that the string has a constant tension. Here, we have an analogue that serves a dual purpose. On the one hand, we have a pendulum, which may oscillate as a pendulum under the force of gravity. On the other hand, we have a string that resembles a double pendulum, because both ends are held fast (though only one end is actually fastened). If the string is displaced but the bob remains in its equilibrium position, it will vibrate under the force of elasticity like a musical string fastened at both ends [*Figure 2.4*].

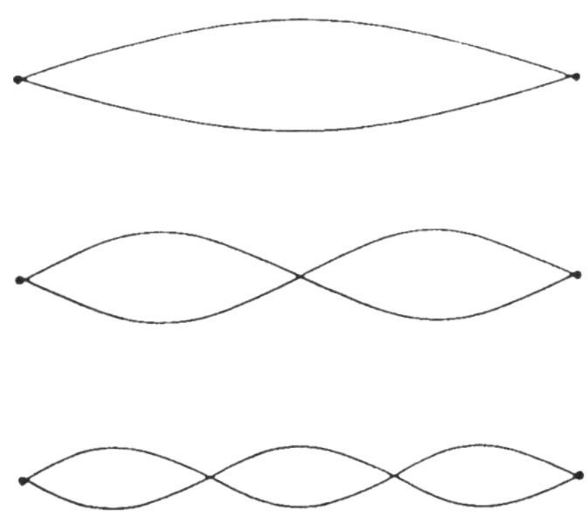

Double pendulum
[First three vibration modes of a string fastened at both ends]

Figure 2.4

Now imagine, with North, a viol string stretched between the nut and bridge, which determine the speaking length (scale) of the string.[93] Here we

[93] The bridge does not function to determine the length (scale) of the string; rather it serves to transmit the strings' vibrations to the belly of the instrument, where the soundpost then transmits the belly vibrations to the back of the instrument.

have an instance of a musical string as a double pendulum moving upon two centres, the nut (upper weight) and the bridge (lower weight), vibrating with the greatest range in the middle of its length. Or, to put it in a different way, here we have one string but two systems with equilibrium configurations along axes, each having the properties of isochronism and simultaneous crossing of the axis.[94] Thus we may understand North, when he writes (p. 18): 'This crossing of vibrations, if the string be true, will be upon equall parts, and produce an octave[,] fifth, or some other chord'—that is, it will produce one of the secondary tones above the fundamental tone [see *Figure 2.4*]. These are the philosopher's harmonics; but to persuade the reader of their existence, North also describes the musician's harmonics—the flageolet tones ('whistle', 'shrill notes') one hears from a bowed string touched lightly at an aliquot part.[95]

He then introduces an instrument capable of producing such sounds successively. This is the trumpet marine [see *Figure 2.5*], a bowed chordophone with a single string about six feet long, stretched over a resonating cavity. To reinforce its sounds, the instrument has an asymmetric bridge, one foot of which is held firmly, the other foot left free to tremble against the soundbox. This is what North describes as the 'jarr at the bridge', which reinforces the sound by vibrating synchronously with the periodic support (the driving force of the bow), thus making 'the sound loud, in imitation of a trumpet'. North, however, does not mention the reinforcement of the sound from strings resonating inside the cavity, a feature that may have been a trade secret. For example, when, on 24 October 1667, Samuel Pepys attended a performance on the trumpet marine, he heard 'a whole consort of chords together at the end of a pause'. On inquiring about this phenomenon, the performer explained that 'it was only when the last notes were fifths or thirds one to another, and then their sounds like an Echo did last, so as they seemed to sound all together'; but this explanation did not satisfy Pepys, who had noticed that although the instrument is open at one end, the performer 'would not let me look into it'.[96]

---

94 See Cannon and Dostrovsky, *The Evolution of Dynamics*, pp. 3–8, who state that until the eighteenth century it was assumed that isochronism meant the restriction to the case of small vibrations.

95 If the string were plucked, one would hear the 'true tone' or fundamental frequency.

96 Pepys, *The Diary*, vol. 8, p. 500. The performer and, apparently, also the maker was the Huguenot 'Monsieur Prin', a London bookseller and father of Jean-Baptiste Prin, a virtuoso performer on the instrument. Both may have been involved in a concert of four trumpets marine, advertised in the *London Gazette* for 4 February 1674; see Tilmouth, *A Calendar*, p. 3.

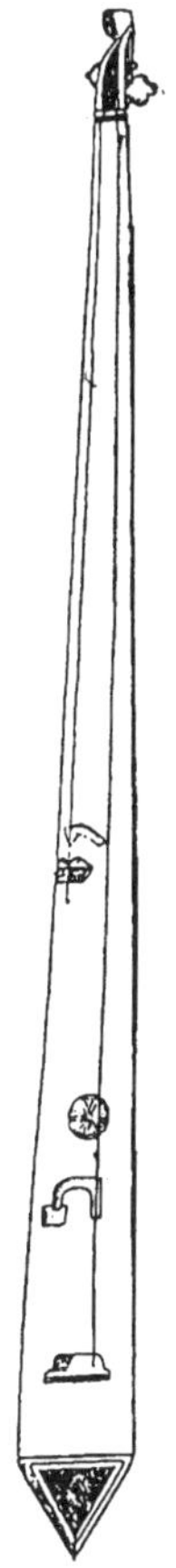

Trumpet marine
[After Mersenne, *Harmonie universelle*, p. 277]

Figure 2.5

Unlike other bowed instruments, where a string is pressed firmly against the fingerboard to produce a vibrating segment of any chosen length, the string of the trumpet marine is touched lightly at the nodal points by the thumb of the left hand, while the right hand bows the string above.[97] Instead of a continuum of pitches, the trumpet marine sounds only a discrete set of pitches that correspond to its natural modes of vibration.[98] An understanding of this instrument, therefore, required an understanding of the connection between nodes—points of absolute or comparative rest in a vibrating body or surface—and higher vibration modes—segmental vibrations of the string that produce secondary tones above the fundamental tone, the vibration of the string as a whole.

Before 1677, it seems that Mersenne alone had investigated the trumpet marine, having learned about it in 1634.[99] Although he never understood the instrument, he drew attention to the fact that its relative pitches are the same as those of a string's flageolet tones, as well as the same as those of the natural, valveless trumpet's overblown tones. He also encouraged others to discover how the instrument operates. North takes two steps in this direction. First, he grasps (p. 19) that for its flageolet tones to sound, the string must be touched on the node ('point'). Second, he grasps that this 'counterfeit trumpet', as well as the 'true' natural trumpet, produce their notes 'by way of breaking', that is, by segmental vibrations. And this latter statement enables him to return to the problem of synchronous vibrations in aerophones (pp. 20–21), where the air column and the nature of its vibrations cannot be easily seen, for, having put his case, step by step, he now can state as highly probable (p. 21) that

> when the note breaks, the column of included air divides into proportionable parts, by reason that the vibrating force at the mouth, and the resistance at the aperture[,] meet one another in the middle, or upon other divisions, which will make the vibrations of the included air coincident with every pulse at the mouth of the pipe. Which is the more probable, because a very small aperture in the middle of a pipe will make it break into an eighth above, which seems not to happen by diminution of the included air, but [by] interrupting the intireness of the column.

In short, just as a stretched string has segmental vibrations, so too does a column of air in a pipe.

---

[97] For details, see Green, *The Harmonic Series*, p. 77.

[98] It is capable of sounding all the pitches of the harmonic series through the sixteenth partial.

[99] Mersenne, *Harmonie universelle*, pp. 276–281; see also Dostrovsky, 'Early Vibration Theory', pp. 194–196, and Green, *The Harmonic Series*, pp. 86–90, 357–361.

Scholars commonly attribute to John Wallis the first experimental demonstration that a stretched string can vibrate simultaneously in equal parts.[100] But this attribution is incorrect, because Wallis's demonstration appeared in a report published in No. 134 of the Royal Society's *Transactions*, dated 23 April 1677, whereas North's *Essay* had been published two months earlier. Moreover, as Wallis himself pointed out, his demonstration was based on the discovery of nodal phenomena 'about three years since'

> by Mr. *William Noble*, a Master of Arts of *Merton-Colledge*; and by him shewed to some of our Musicians ... ; and after him by Mr. *Thomas Pigot*, a Batchelour of Arts, and Fellow of *Wadham-Colledge*, who, giving notice of it to some others, found, that (unknown to him) the same had been formerly taken notice of by Mr. *Noble*, and (upon notice from him) by others: and it is now commonly known to our Musicians here [in Oxford].[101]

To make this discovery visible, Wallis placed small rings of paper, now called 'paper riders', around a vibrating string. When the string vibrated, the riders shook vigorously almost everywhere along the string but remained more or less at rest at the nodes ('points').

Wallis's demonstration relied on synchronous vibrations in strings that were not tuned in unison. First, he took two strings stretched side by side and tuned an octave apart. He found that exciting the higher pitched string caused the lower one to resonate and to divide into two equal sections separated by a stationary point. Next, he tuned the two strings a twelfth apart. In exciting the higher pitched string, the lower one divided into three sections with two places at which the paper riders were motionless. Then, when the strings were tuned a fifth apart, exciting the higher string caused the lower one to divide into three, whereas exciting the lower string caused the higher one to divide into two. He also made a similar trial with other sections. From the observations, Wallis concluded not only that a vibrating string breaks into equal vibrating segments ('unison parts') but also that a similar situation must exist in aerophones. Thus (perhaps), when Pepys purchased a copy of North's *Essay*, he assumed that Wallis was its author.[102]

---

[100] See, e.g., Dostrovsky, 'Early Vibration Theory', p. 204. For a critical appraisal of Wallis's report, see Green, *The Harmonic Series*, pp. 380–383.

[101] Wallis, 'Dr. Wallis's Letter to the Publisher', pp. 840–841. Later, Wallis stated that he heard about the discovery in 1676 from Narcissus Marsh; see Wallis, 'De algebra tractatus', *Opera mathematica*, vol. 2, p. 466. (The treatise is an augmented Latin translation of an earlier English version, where no such statement occurs.)

[102] Pepys, *The Diary*, vol. 10, pp. 278, 536.

Wallis, who seems to assume that string length is the only important factor that controls timing and determines pitch, never mentions the phenomena of fundamental and secondary tones, whereas North makes explicit reference to these phenomena. And Wallis never explicitly asserts that higher modes of vibration sound simultaneously with the fundamental. North does not explicitly state this either, although he implies it, when (pp. 17–18) he writes of a 'single sound' as a compound produced by a composition of synchronous vibrations,[103] and when he provides a *visible* demonstration of this in two graphs (*Figure 1.A*, Tab. i and ii). It is remarkable, therefore, that on 15 January 1676, a year before the publication of North's *Essay*, Hooke seems to have devised an *audible* demonstration, as we learn from an entry in his daily diary. Since the entry is so important, I will quote it in full:

> To Sir Chr. Wrens, Dr. Holder and I discoursd of musick.... I told him but *sub sigillo* my notion of sound, that it was nothing but strokes within a Determinate degree of velocity. I told them how I would make all tunes by strokes of a hammer. Shewd them a knife, a camlet coat, a silk lining. Told them that there was no vibration in a puls of sound, that [sound] twas a puls propagated forward, that the sound in all bodys was the striking of the parts one against the other and not the vibration of the whole. Told them my experiment of the vibrations of a magicall string without sound by symphony [i.e., sympathy] that touching of it which made the internall parts vibrate—caused the sound, that the vibrations of a string [as a whole] were not Isocrone but that the vibration of the particals was. Discoursd about the breaking of the air in pipes, of the musick of scraping of metall, the scraping the teeth of a comb, the turning of a watch wheel &c., made sound. Compard sound and light and shewd how light produce colours in the same way by confounding [i.e., compounding] the pulses.[104]

There are five points to note about this memorandum, which might serve as an outline for the first two parts of North's *Essay*. First, Hooke presented a brief definition of sound (and, in the last sentence, light) as pulse fronts of 'determinate' timings—frequencies. Second, against the belief that a string vibrating in open air causes compressions and rarefactions which are heard as sound, Hooke concluded by experiment that a vibrating string is not the source of sound, since it can be made to vibrate silently and to impart its motion to

[103] FN's statement (pp. 8–9) may refer to mixtures of harmonics that 'produce an alteration'; but he does not recognise that the composite quality is changed when any secondary tone is absent, a recognition that did not appear explicitly until the work of Helmholtz.

[104] Hooke, *The Diary*, p. 211. Throughout this edition, there seem to be a number of errors in transcription, no doubt because Hooke's handwriting is difficult to read.

another string.[105] Third, he hinted that tone ('sound') is a composition of simultaneous motions of the whole string and its aliquot parts ('particles'), each part undergoing simple pendular vibrations having the same period (i.e., the vibration of each segment is isochronous).[106] Fourth, and following this hint, he mentioned the 'breaking of the air in pipes', which implies an analogy between a vibrating air column and a vibrating string. Fifth, he suggested cryptically that the way to get a pitch ('sound') is to get a frequency—by scraping metal[107] or the teeth of a comb and by turning a watch wheel. A few months later, on 28 March 1676, Hooke wrote in his diary: 'Directed Tompion about sound wheels. The number of teeth.'[108]

Hooke's two 1676 references to sound wheels have had little notice, because it has been assumed that he first used them in an experiment performed on 27 July 1681 before fellows of the Royal Society, the minutes of which record that the experiment consisted of making 'musical and other sounds by the help of brass wheels; which teeth were made of equal bigness for musical sounds, but of unequal for vocal [i.e., speech] sounds'.[109] And after Hooke's death, Richard Waller also recorded that Hooke had demonstrated

> a way of making Musical and other Sounds, by the striking of the Teeth of several Brass Wheels, proportionally cut as to their numbers and turned very fast round, in which it was observable, that the equal or proportional stroaks of the Teeth, that is, 2 to 1, 4 to 3, &c. made the

---

[105] See also Hooke, *The Diary*, p. 210, for 5 January 1676: 'I told them [Abraham Hill and Christopher Wren] I could make a string vibrate without giving any sound but told them not how, meaning the sympathetick motion.'

[106] Hooke, *Micrographia*, p. 78, may have had a similar notion about artists' colours, for he held that

> most of the colours ... they use in Limning and Painting, are such, as though mixt with any other of their colours, they preserve their own hue, and by being in such very smal parts dispers'd through the other colour'd bodies, they both, or altogether represent to the eye a *compositum* of all; the eye being unable, by reason of their smalness, to distinguish the peculiarly colour'd particles, but receives them as one intire *compositum*: whereas in many of these, the *Microscope* very easily distinguishes each of the compounding colours distinct....

[107] An idea possibly derived from Galilei, *Two New Sciences*, pp. 101–102, where Galileo reported that a sound of determinate pitch was sometimes produced as a sharp iron chisel was scraped across a smooth copper plate, the scratch leaving a series of equidistant parallel lines that were closer together when the pitch of the sound was higher.

[108] Hooke, *The Diary*, p. 223. The person directed was Thomas Tompion, a clock and mathematical instrument maker; see Brown, 'Guild Organisation and the Instrument-making Trade', p. 33.

[109] Hooke, 'The Life and Work', vol. 7, p. 577.

> Musical Notes, but the unequal stroaks of the Teeth more answer'd the sound of the Voice in speaking.[110]

Neither record provided much detail about the experiment, but we learn considerably more from Roger North, whose manuscripts contain three different descriptions, two of which were written sometime during the period c.1704–10 and the last, sometime between c.1726–33.

In one of the earlier accounts, we find the following:

> The ingenious Mr. Hook, made an engin of wheels that made pulses in any musical proportion, as 2, 3, 4, 5, or 6 to 1 and so 3 to 2 and the like. This latter is that they [musicians] call a fifth in musick. These wheels he would turne first slow, so as the sound of the pulses was distinguishable, and so long nothing extraordinary was observed[;] but as the motion accellerated, and all distinction of pulses was gon[e], then began that harmonious ringing of the musitian's fifth—which in the imagination was a creature of meer confusion, and not otherwise existent. And all those other ways of perceiving, which wee call senses, have the same faculty, even light and colours, are but the result of confused apprehension of numerous pulses on the organ [of sense], as may be more largely discourst elswhere.[111]

In the other account written during the same period, some additional detail is provided:

> I have heard that the ingenious Mr Hook made an instrument of wheels to the movement of which he could order what proportions he thought fitt, and to these wheels he contrived movable coggs that in turning should strike upon a table, or stone, isocronically. And some wheels should strike together puls for puls, and others in proportions, as 1/2, 3/2, etc. and then applying and mixing as he thought fitt. He would begin to turne slow, and so long the pulses were distinct, and he could discerne them, as smiths at anvill, without any other idea; but then coming to a mighty swiftness, the consonance called fifth (for instance,) which is 3/2. He could perceiv, from an idea onely of pulses it fell into[,] that known twang, a new idea, of that a fifth yeilds, and so of others; which sort of demonstration of the nature of musicall accords is irrefragable. And I cannot but wonder yet, that this impression of acceptance or disgust should be so egreggious in its kind, as wee have from mixt sounds that

---

110 Richard Waller in Hooke, *The Posthumous Works*, p. xxiii.

111 See ff. 33–33v in RN's essay, 'The world' (GB:Lbl) Add MS 32546: ff. 33–90, which is bound in a notebook with other essays, for which see *NP 4*, pp. 86–87. For the matters to be 'more largely discourst elswhere', see also *NP 5* (c.1704–c.1709).

> proceed from pure stroking onely without interposition of the understanding, or the least information of the nature of them.[112]

Note that, in both descriptions, Hooke's engine is said to have comprised several wheels for demonstrating the characteristic periodicities of intervals.

Of special importance, however, is the account written between c.1726–33, because Roger North specifically states that Hooke converted his brother's 'Scheme' into clockwork,

> and made wheels, with small lingulæ, in the manner of coggs, which moving each upon its pinn, as the wheel turned, struck upon an edg, one after another equably; the wheel turning slow the pulses were distinguishable, and had no other vertuc; but then turning swifter, the distinction ceased, and a plain musicall tone emerged. This for one[;] then, another wheel was contrived to strike 3 to 2 (for instance) and as the distinction begun to fail, and continuation took place, one might hear a consort 5th coming on, and setling in the manifest accord so named.[113]

In the second description quoted above, Roger North asserted that his source was hearsay: 'I have heard.' Although the informant could have been Francis North himself, other sources might have been Hooke or Wren, since on 8 December 1688 Hooke recorded that he met and discoursed with Wren, Dudley North and his 'brother', that is, Roger North.[114] If this was the occasion when Roger North learned about the experiment, then perhaps Hooke did indeed devise it in 1676 to provide an aural demonstration of Francis North's 'Scheme'.

Of the numerous scholars who have commented on the 1681 experiment, Clifford Truesdell merely observed that Hooke had 'devised an experiment for producing sounds by toothed wheels, (but what he did is hard to ascertain)'.[115] Burdette Green supposed that Hooke failed to experiment with 'overtone relationships', because he was unaware of 'the similarity between their pattern and North's representation' (*Figure 1.A*).[116] Against this supposition, the evidence suggests not only that Hooke saw sheets of North's *Essay* as they

---

[112] See ff. 91v–92 in RN's untitled treatise, (GB:Lbl) Add MS 32537: ff. 66–109 (a revision of *Cursory Notes of Musicke*); for a description of this manuscript, with an analytical index of it, see *NP 5*, pp. 49–170.

[113] RN, *Life/FN*, p. 250.

[114] According to Hooke, 'The Diary', p. 80, the conversation concerned 'Turky building, Pottery, Humhums etc.'—the Hummums in Covent Garden being a bagnio or place for sweating. For RN's contact with Wren, see Chapter 3, p. 112 n.95 *infra*.

[115] Truesdell, *The Rational Mechanics*, pp. 53–58.

[116] Green, *The Harmonic Series*, p. 395.

came from the press but also that Hooke had a number of wheels for exploring the distribution of sound energy among different frequencies as a function of time. Other scholars have regarded the experiment as demonstrating the correlation of pitch and frequency. Frederick Hunt, for example, cited the experiment as a precursor of one made by Felix Savart early in the nineteenth century,[117] and Sigalia Dostrovsky concluded that

> Hooke's wheel demonstrated experimentally, essentially for the first time, that which Galileo had demonstrated conceptually, namely, the correctness of identifying pitch with frequency or, more precisely, interval with relative frequency. Furthermore, it provided a natural and obvious means of determining absolute frequency. But the accuracy of the method would be limited, in particular, by the fact that the wheel produces a very impure tone.[118]

Certainly, the method would have been less accurate if the wheel had struck a stone, as Roger North claims. But the tone could have been produced in a different way, namely, by a watch spring attached at one end and applied to the teeth at the other end. According to this method, the spring, being raised by each tooth as it passes, would produce a tone having its frequency (pitch) number equal to the number of teeth by which it is struck in a second.[119]

In whatever way the tone was produced, Hooke's experiment would have given a more accurate estimate of the rates of vibration of particular pitches than would the earlier experiments of Mersenne, who had scaled down the frequency by working with very long cords that would vibrate slowly enough to make possible the direct counting of vibrations. As a number of subsequent investigators admitted, counting vibrations is a tedious business. Moreover, results of investigations depended on what is counted. For example, some experimenters counted the forward and backward motion of a vibrating body as one vibration, whereas others counted the forward motion of a vibrating body as one vibration and the backward motion as another, so that the whole vibration was counted as two.[120]

---

117 Hunt, *Origins in Acoustics*, p. 136.

118 Dostrovsky, 'Early Vibration Theory', p. 199.

119 A careful exploration of Hooke's work in horology might reveal more about the nature of his 'sound wheels'. From the 1660s, for example, he experimented with springs as regulators for watches; and on 18 March 1673 he began work on his wheel-cutting engine, which the editor of his papers described as 'an Engine to cut down the Teeth of Watch Wheels more exactly than can be done by the most expert Hand, an Invention now of constant use'. See Hooke, 'A Description of Helioscopes', pp. 146–152; Hooke, *The Diary*, p. 34; Richard Waller in Hooke, *The Posthumous Works*, p. ix.

120 For Mersenne's attempts to count frequencies, see Hunt, *Origins in Acoustics*, pp. 89–95; for others' attempts, see Helmholtz, *On the Sensations of Tone*.

It is important to remember that Hooke's experiment was devised during a period when there was considerable debate about the nature of sound. Understood from this point of view, his experiment offered proof of the kinematic variant of the continuum hypothesis—that sound is a motion (pulse, vibration) that takes time. Using unequal- and equal-toothed wheels for noise and tone, respectively, he showed that, as the rate of the motion of the wheels increases, a noise or tone emerges that seems continuous, because the pulses are too rapid to be counted. This is the central point in Roger North's descriptions of the experiment, perhaps because the experiment supported his own hypothesis, to be examined later on, concerning the origin of musical knowledge.[121] It also is important to remember that, if Hooke devised the experiment in 1676 to demonstrate Francis North's new definition of tone as a compound or sum of simple pendular vibrations, then the problem of *absolute* frequency ceases to be a problem. Although physicists require such absolutes for scientific purposes, their purely theoretical or 'philosophical pitch' is not a requirement for musical purposes, where slight defects in pitch, tempo and other properties give 'life and warmth' to music.[122]

Insofar as pitch is the property concerned, very slight discrepancies produce slow beats, that is, long drawn out fluctuations alternately swelling and vanishing that are easily countable. As Helmholtz later pointed out, such beats are not disagreeable to the ear and may be used in performance to give expression to the music.[123] And Roger North voiced a similar (though more eccentric) opinion about defects in pitch and tempo, when he wrote:

> It is a strang[e] affirmation, but I will venture upon it, being now ... adventrous ..., that consort musick is better when not exquisitely performed, then otherwise. For if it be nicely done with severall instruments to a part (as must be in loud musick) you have onely the harmony designed, which may be for the most part ordinary, and all is as one instrument invigorated [i.e., reinforced by resonance]. But if there be errors of performance intersperst[,] it hath the force of artificiall discord, as one part sometimes moving before or after another, and not pitching exactly on the same notes, but varying a litle, gives an excellence few have observed or will allow. But this must be with great moderation and temper for it must have that effect, of affected discords, to be a litle and not much out of the way, and so season and sparkle in the accord as air and light doth to the eye, in a landscape. And [it] is an elegance, which painters in their art cannot describe and musitians must not thinck to

---

[121] See Chapter 3, pp. 117–120 *infra*.

[122] See RN, *Notes of Me*, p. 158; see also Winckel, *Music, Sound and Sensation*, p. 133: '"Hairbreadth" accuracy ... is perceived as dull or mechanical.'

[123] See Helmholtz, *On the Sensations of Tone*, p. 167.

> carracterize their perfections; for in all arts, the soveraigne beautys as of woemen are *je ne say quoy.*[124]

Later, he developed his brother's hint—that beats are the cause of dissonance (pp. 10–11)—by imagining that the regularity in beats of discordant sounds might be made use of in musical practice.[125]

THIRD PART: THE SEMANTICS OF MUSIC
'OF THE NATURE OF KEYS IN MUSICK, AND OF A SINGLE TUNE' (pp. 21–24)

In the third and last part of his *Essay*, North turns to the meanings that the mind makes on the basis of sense perceptions, so that we now move from the mechanics to the semantics of music. In his first sentence, he also makes a subtle distinction between the terms 'tone' (a mechanical term) and 'note' (a musical term) as follows: 'Before any note [is] struck[,] the ear is *indifferent* to all sounds, but when any tone is heard, that *indifferency is determined*, and the mind [is] taken up with the present sound' (italics mine). To understand the import of this sentence, recall North's two statements about musical instruments as coupled systems of acoustic vibrators and resonators (p. 16): first, that their elastic force ('spring') stands '*indifferent* to take any measure' until the timings of 'the vibrations are *determined*'; and, second, that their 'tremble' 'changes with every new sound' (italics mine).

Although North there mentions 'the nerve of hearing', he has little else to say about the anatomy or physiology of the ear.[126] Nevertheless, from his two statements we may infer, first, that, like a musical instrument, the ear itself is a coupled system; second, that the theory of hearing implied in the *Essay* is a resonance theory; and third, that when a note is 'struck', what the ear hears subconsciously is a simultaneous set of fundamental and secondary tones, like the major chord over a fundamental bass depicted in *Figure 1.B* of North's 'Scheme'.[127] These inferences lead to the conclusion that the ear is the

---

[124] See f. 4v in RN, 'Some Memorandums, concerning Musick, (GB:Lbl) Add MS 32532: ff. 1–26v and *NP 3*, pp. 36–31.

[125] For the first development, see RN, 'The Theory of Sounds...', (GB:Lbl) Add MS 32534: f. 43v, for its development; see *NP 3*, p. 94; see also Chenette, *Music Theory*, p. 205.

[126] Hobbes, 'Elements of Philosophy', p. 404, assumed that the 'proper organ of hearing is the tympanum of the ear and its own nerve'. In 1660 Thomas Willis, physician to FN's wife, published an improved anatomy; see RN/*Life FN*, p. 213.

[127] For RN's interpretation of his brother's two statements, see f. 13v in RN, 'Some Memorandums, concerning Musick', (GB:Lbl) Add MS 32532: ff. 1–26v, where we find the following:

connecting link between the mechanical events in the external world and the meanings the mind makes on the basis of sensation.

At the auditory system's subconscious level—the level of sensory perception, the periodicities are too fast for conscious counting. But at its conscious level—the level of perceptual cognition, the periodicities are slower, so that, in treating this level, North can apply Hobbes's temporal framework for conception that involves differentiating the past (remembrance) from the future (expectation or anticipation) by asserting the present (immediate sensation).[128] In this application, North begins with music in one part only—a 'single tune', in which the composer affects a purely internal musical order among all the notes by making their relation to one note—the keynote—as perceptible as possible. It is this relationship that provides the attentive listener with a feeling of the tune's coherence, for without such a relationship 'the sounds' would seem 'disjoynted'.

Thus (pp. 22–23), 'in the progress of a tune[,] the key is to be traced by the relatives to it, as well as by the [key] note itself'. To grasp the meaning of a tune, the attention oscillates between the succeeding and the preceding notes through the centre of oscillation, the keynote, until brought to rest. If the notes have a close relation to the keynote, they will strengthen the remembrance of the keynote. If other notes have a less close relation, the first keynote may be forgotten and the ear may be determined to a new keynote. If the fifth 'comes into play', this note, which has the closest relation, not only strengthens the remembrance of that keynote but also increases expectation of its return.

Since the keynote determines the ear, it also connects all the notes in a piece of music into a coherent order. This is the pre-compositional principle that we call 'tonality' and that North terms 'formality' (p. 23), perhaps taking his cue from Charles Butler, who, in 1636 (using his own phonetic system), had defined 'formality' as 'de mainteining of de ... Air, or Ton' of de Song, in his Part's'.[129] Here, however, North restricts formality to 'a single part', and he offers a number of suggestions for achieving it. One way is to begin and end on the keynote, although he acknowledges that this is not a hard and fast rule, especially in longer tunes. Hence, he suggests (p. 22) that formality is best

---

> [The] full accord is that which nature makes and not art nor invention. And although the ear is indifferent to all sounds, or tones, yet if one tone be sounded[,] that indifference is determined. For the accord is so vertually included in it that wee will not endure to hear, that to suceed which is bad discord. For tho the sound be ceast[,] the memory of it, remains and the following notes, and the memory of the first, are connected, and must be harmony or all is naught.

See also f. 43 in 'The Theory of Sounds...' (GB:Lbl) Add MS 32534: ff. 1–82v, where RN states: 'when a musicall sence hears a tone, it partly hears all its accords'.

128 See Chapter 1, pp. 14–15 *supra*.

129 Butler, *The Principles of Musik*, p. 81; see also pp. 44, 57 (=55), 62, 81–88.

achieved by repeating with 'length', 'loudness' and 'emphasis' not only the keynote but also its closest 'relatives', the fifth and third.[130] Repetition of these 'relatives' strengthens the remembrance of the keynote, so that in the final cadence of a tune, when the fifth 'comes into play', expectation is satisfied by the return to the keynote at the conclusion of the tune, for this gives 'a perfect acquiescence', or feeling of rest, to the listener.[131]

In antiquity, a similar approach was identified by Aristoxenus, who taught that 'musical cognition implies the simultaneous cognition of a permanent and of a changeable element'; that this implication 'applies without limitation or qualification to every branch of music'; and that 'in matters of harmony' it is necessary to accustom 'both ear and intellect to a correct judgement of the permanent and changeable element alike'.[132] Although the system of music in antiquity differed from the 'systeme' North calls 'formality', Aristoxenus took an empirical approach to perceptual cognition, and it is this approach that Hobbes and North both share.

'Of Schismes, and the Scale of Musick' (pp. 24–28)

North next endeavours to show (p. 24) 'how all the notes come into the [diatonic] scale by their relation and dignities', so that it will be obvious to the reader why 'the scale of degrees of musick is made as the musicians now exhibit it'. North's chief concern, however, is to address a particular problem that arises from modulation—the process of changing from one key to another during the course of a musical composition. As Carlton Maley pointed out,

> Modulation is carried out by taking one pitch in two senses; this pitch will have one relation to the tonic note [the keynote], or 'key' of the music which precedes it and another relation to the tonic of the music which follows it. [Hence] ... this pivotal pitch must have one and the same frequency while it stands in different relations to the two tonics. But every modulation from one tonality to another requires the introduction of one or more new pitches. As this process is continued, eventually a tonality will be reached which will require a note quite near one which has been introduced earlier.[133]

---

[130] FN's 'length', 'loudness' and 'emphasis' are, respectively, quantity, stress and accent—what linguists now term 'suprasegmentals'.

[131] See Herissone, *Music Theory*, pp. 183–184, for the importance of FN's treatment of the fifth as a secondary centre beyond the keynote.

[132] Aristoxenus, *The Harmonics*, pp. 189–190.

[133] Maley, *The Theory of Beats*, pp. 140–141; see also Cohen, *Quantifying Music*, pp. 215–217.

The requirement of two distinct pitches of nearly identical frequency poses no problem for the human voice or for unfretted instruments like the violin, since these are capable of producing almost any slight variation in pitch that may be required. But the case is different with fretted and with keyboard instruments, which produce only a predetermined number of fixed pitches, one for each fret or key, so that the requirement of separate pitches in these cases might be resolved in one of two ways: either by modifying the instrument or by using some kind of temperament.[134]

Previously, North alluded to this latter case, when he wrote (p. 23) that there is 'no natural difference betwixt the music in one key note or another', although there are accidental differences due to instruments with fixed pitches, where the 'emphatical notes'—the keynote, fifth and third—'are not in tune'. He now treats both methods of resolving the problem of modulation for instruments with fixed pitches, beginning with temperament. This term, applied to a musical interval, means the slight alteration of that interval from an exact Pythagorean or other whole number ratio in order to meet the necessities of musical practice. Initially, North (p. 25) gives a practical method for tuning intervals within the compass of an octave; but, as he points out, this tuning is 'exact' for music in one key only. When music modulates to other keys, there will be variations in frequency of two nearly identical pitches, as noted above, so that temperament is necessary—here, a variety of meantone temperament,[135] which North instances in the division of the monochord (p. 26 reproduced as *Figure 1.C*) and treats briefly in the accompanying text (p. 27).[136]

North has a special name for the variations in frequency of two nearly identical pitches that arise during the course of modulation—he calls them 'schismes'. His usage, therefore, is not to be confused with that of Descartes, for whom schism denoted the difference between a major tone and a minor

---

[134] Fretted instruments were usually tuned in equal temperament, whereas keyboard instruments were usually tuned in meantone temperament, although with the latter variety of temperament keys with more than about three flats or sharps were unusable. Nevertheless, meantone was the common temperament for organs as late as 1851, according to Alexander Ellis in Helmholtz, *On the Sensations of Tone*, p. 434.

[135] Of the different methods of achieving meantone temperament, the most common was by tempering fifths in such a way that all the other intervals were incommensurable. In other varieties, both fifths and thirds were altered; and in still others, the chosen intervals—the fifths, or the fifths and thirds—were adjusted either uniformly or by making some intervals bear more than others. See Barbour, *Tuning and Temperament*, pp. 25–44, 133–151; see also Messen, 'A Question of Temperament'.

[136] Since the printer introduced errors into FN's division, *Figure 1.C* is reconstructed in a clearer format with errors corrected; see Appendix A, *Figure 1.D*, pp. 167–170 *infra*.

tone, the so-called syntonic comma.[137] This usage reflects Descartes' reliance on arithmetical proportions to produce a diatonic scale in just intonation.[138] In his animadversions, Brouncker proposed two other methods for accomplishing scale division, the first, algebraic, and the second, by the use of logarithms—including, for the first time, a logarithmically determined equal temperament.[139] In so doing, he undermined the validity of arithmetical proportions; and he also challenged Descartes' representation of music as visible line lengths, when he wrote:

> Now considering (as was sayd An[imadversion]. 1 and An. 3) that not the *visible* proportion of *Chords* or *Strings*, but the *audible* proportion of their *Sounds* only is considerable in Musick; and that, by the *Sence* of *Hearing*, wee doe judge of *Sounds* according to the *Geometricall*, not Arithmetical Proportion, or proportionall *Division* of the *Strings*, that give them: I conceive it was rightly inferred An. 3, that *Chordes*, as to Sounds, ought to bee divided according to a *Geometricall*, not Arithmeticall *Progression*; by force of the same *Reason* (adequated to the Sence of Hearing) which our *Authour* [Descartes] gave for the contrary opinion....[140]

Against the rationalism of Descartes, therefore, Brouncker stressed that sense is the judge in the manner of scale division, for 'the *Sense* of *Hearing* is not so perfect, as to confine the *Consonances* to so precise a *Measure*'.[141] North echoes these sentiments, when he asserts (p. 24): 'the degrees in the scale of musick are fixed by the ear in those places, where the pulses of the tones are coincident, *without any regard to the quantity*' (italics mine).

As noted above, there was a second way in which the requirement of separate pitches might be resolved, namely, by modifying instruments with fixed pitches. In England, Christopher Simpson had described modifications in relation to both fretted and keyboard instruments.[142] Even though North himself was a performer on a fretted instrument, he restricts his brief remarks

---

[137] For Descartes' usage, see *schism*: OED *sb.* 4 = *comma* OED 3 (1) *obs.* The OED does not record FN's usage, which, subsequently, was adopted by RN; see, e.g., *NP 2*, p. 27, and *Cursory Notes of Musicke*, pp. 166–173.

[138] Just intonation is a system of tuning based on the octave (2:1), the pure fifth (3:2) and the pure major third (5:4).

[139] For details and also for the erroneous description of Brouncker's system in the prefatory 'Stationer to the Reader', see Chenette, *Music Theory*, pp. 128–134, 170–171.

[140] Brouncker in Descartes, *Excellent Compendium*, p. 84; for Animadversions 1 and 3, see pp. 62 and 63.

[141] Brouncker in Descartes, *Excellent Compendium*, p. 89.

[142] Simpson, *A Compendium*, pp. 78–79.

(pp. 27–28) to the modification of keyboard instruments by 'quarter notes'—that is, additional or split keys—placed where the 'schismes' are the greatest, namely, the enharmonic dieses that produce the effect traditionally compared to the howling of wolves.[143] Neither Simpson nor North indicate how these extra keys are to be tuned, although J. Murray Barbour 'freely' assumed that regardless of the variety of meantone temperament used 'for the twelve regular notes', it was extended to the extra keys 'both clockwise and counterclockwise around the circle (or, rather, spiral) of fifths'.[144]

Before leaving this section of the *Essay*, one other feature deserves notice. This is the illustration at the bottom of 'A Monochord' (*Figure 1.C*) of what North terms a flat and a sharp key—A and C—or what we would term the relative major and minor mode. From the second decade of the seventeenth century, the tendency in writings on musical practice was to reduce the eight or twelve church modes to two, so that the question arises: which two? As Walter Atcherson pointed out, there were three different answers to this question, only one of which fits our modern concept of relative major/minor, namely, the pairing of C major with A minor. North's illustration is an early—perhaps the earliest—instance of this pairing, since Atcherson assumed that it appeared for the first time in a tract published in 1688 with the rationale that a viol player could play in those two keys without having to change fingerboards.[145]

'OF MUSICK CONSISTING OF SEVERAL PARTS IN CONSORT' (pp. 28–33)

In treating music in several parts, North declines (p. 32) to describe 'excellent musick' since that 'would require a poet as well as a musician'; rather, his aim is that of a philosopher—'to show what is allowable musick, in order to make the reasons[,] upon which their rules are founded[,] understood'. He therefore reminds the reader (p. 28) 'that tones in chord one to another will joyn and give one entire sound, whereas discording tones will jar'—that is, beat; and also that

---

143 This was a consequence of English organs, which used only twelve notes to the octave; see Alexander Ellis in Helmholtz, *On the Sensations of Tone*, pp. 434, 551–552. After 1677 split keys were incorporated in organs that Smith built for the Temple Church (1683, purchased 21 June 1688) and Durham Cathedral (1684). The former organ has been described as the first organ in England to have the additional keys for A♭ and D♯ distinct from G♯ and E♭; see McCrory, *A Few Notes on the Temple Organ*, pp. 33–35. RN, *Cursory Notes of Musicke*, p. 173, allowed that split keys 'accomodate any [keyboard] instrument to ordinary use'; but he seems to have been unconvinced by such contrivances: 'after all allowing as many plaisters as you will, there will be sores'.

144 Barbour, *Tuning and Temperament*, pp. 107–108 (whose chief interest seems to have been the use of multiple division to represent just intonation).

145 See Atcherson, 'Key and Mode', pp. 225–226, who incorrectly referred to FN as 'Lord North' and gave an erroneous date for the *Essay*.

'some chords were more perfect than other[s], according as the coincidence of their pulses was more frequent'. This reminder refers to the 'Scheme' (pp. 2–3) and its related text (pp. 7–11), where the principle of relative perfection of consonances by the coincidence of pulses observed in the harmonic series constitutes the rationale for North's principle of formality—the internal order and relation of consonant notes to a keynote.[146] According to North (p. 30), formality is a principle that 'every part ought to have', so that this principle includes both 'harmony' (p. 28) and 'conformity' (p. 31). Harmony is 'the gratefull sound produced by the joyning of several tones in chord one to another', whereas conformity is the balance or symmetrical relation of all parts of a composition. Taken in this extended sense, the principle of formality serves to guide the composer, and, when well applied, to provide the attentive listener with musical meanings.

Prior to 1677 both philosophers and writers on the practice of music ordered individual consonances categorically, without consideration of their function in a musical context.[147] North, however, adopts a different approach, since his principle of relative perfection provides a rationale for the importance of the bass, the identity of octaves and the function of 'chords'—that is, consonant notes—above a fundamental bass. The importance of the bass appears from *Figure 1.B* of North's 'Scheme' (p. 3), where the fundamental tone is the norm with which the pulses coincide and which, in sensory perception, we hear as a simultaneous set of secondary tones present in a fundamental tone. But North now points out (pp. 29–30) that, 'according to nature', the pulses of the fundamental tone are 'bigger and stronger' than the secondary tones, so that they are of more effect in maintaining the coincidence.[148] Hence, in music consisting of several parts, the bass also is 'the foundation, and all the upper chords are gratefull as they have relation to that, because it comprehends all the rest'. Furthermore, the relationship of the 'upper chords' to the bass is determined by the order of the secondary tones to the fundamental tone.

As for the identity of octaves, North writes (p. 29) that

> as in division [multiple notes in the time of one], the mode of the time is never alter'd by doubling the number of notes in the same space of time,

[146] From about 1698 onwards, RN uses this principle of relative perfection to justify harmonic structures and to form a system of harmonic progression; see Chenette, *Music Theory*, p. 277.

[147] For the philosophers, see Cohen, *Quantifying Music*; for the writers, see Chenette, *Music Theory*, and Herissone, *Music Theory*.

[148] According to Helmholtz, *On the Sensations of Tone*, p. 57, the fundamental or 'prime tone is generally louder than any of the upper partial tones, and hence it alone generally determines the pitch of the compound'; but, p. 58, sometimes 'the loudness of the first upper partials is not inferior to that of the prime tone itself'.

> but the mind or memory couples them and comprehends them in the same space [of time] as if they were whole notes; so is it by octaves, being [that] they increase by duplication of the pulses, in the same space of time [so that] it [duplication] produces no alteration of the sound.

In short, although octaves differ from unisons, 'yet as to appearance, they are the same, in respect of such differences as there is between other sounds'. Like unisons, therefore, octaves do not function as consonant notes; rather, their purpose is to fill the harmony.[149]

North does not develop a system of harmonic progression,[150] since he restricts his comments on consonant notes above a bass to the 'close', the cadential passage that concludes a piece of music. In the context of closes, the order of perfection depends on whether consonant notes provide a stable harmony above the bass ('the 'foundation') or whether they induce change to another key. For example, since the fifth and major third have the closest relation to the keynote, they are the principal consonant notes, the function of which is to provide the stability in which the mind acquiesces. Because the fourth is unstable and induces the mind to expect change, it functions as a binding or syncopated note, not as a consonant note.[151] The minor third and the major and minor sixths have no place in the close, 'because they are in discord to the third sharp and fifth'; but in 'the body of a tune', these notes provide stability and, hence, 'never offend the ear'.

Although discords are not allowed in music 'for their own sake', they are in other respects of 'excellent use', as, for example (pp. 31–32),

---

149 The octave has all its harmonics coincident with upper harmonics of the fundamental tone. No other musical interval has this property (except the unison). For the importance of this property in pattern recognition, see Roederer, *Introduction to the Physics and Psychophysics of Music*, pp. 147, 152–153.

150 Chenette, *Music Theory*, p. 370, identified three stages in the development of harmonic structures. Data for the first stage are to be found in treatises written prior to 1660, where harmonic sonorities are created by combinations of melody lines according to categorical rules for use of specific intervals. Data for the second stage stem from early thorough-bass treatises, where intervallic sonorities are accepted as entities with harmonic meanings that are not necessarily implied from melody. RN, developing hints from FN (see p. 82 and n.146 *supra*), initiates the third stage, in which there is awareness that a combination of pitches has some sound as a tonal focus, that this sound is not necessarily the lowest written note and that the tonal focus, with its constituent pitches, has a relationship to a key centre.

151 See f. 8v in RN, 'Some Memorandums, concerning Musick', (GB:Lbl) Add MS 32532: ff.1–26v, where, in writing about 'softning and loudning', he states:

> Binding notes or emphaticall discords (as they are esteemed) ... which usually in practice, are continued over into a new time or barr (as the wrighting usually is) ... should be prest hard, for the elegance and surprise excites a great attention.... And then when you come off into a sweeter[,] calmer air as to cadence, which often follows such passages, then be soft and easy, as much as to say, be content[,] all is well.

> when the mind is to be carried from any particular note, it may be affronted by a discord; or if a note in small esteem be to be favoured, some other note may come in that agrees with that, and is discord to the rest, and so may draw the mind to the regard of that note which it strengthens and supports.

As a general rule, however, the 'sweetness' must be distributed amongst the parts, even though, in some instances, 'the present harmony' must yield, for (p. 31)

> if the succeeding notes are of great relation to the key, and the formal way to them in the several parts is by discording notes; nevertheless that way which gives so good expectation must not be balked, but rather the discords will be allowed, whose harshness is taken off by the expectation of the succeeding chords and the harmony of the following notes is the more welcome for the badness of the passage.

Finally, there is conformity, whereby the 'contexture' of the parts should be 'fairly interwoven', so that one part does not 'cling to another' or move, parallel, in consecutive fifths and octaves, for this would 'draw the attention too much'.[152] To avoid this, there must be order, connection and symmetry of all the parts and internal relations, so that (p. 32) musical coherence occurs

> when each part has its kind and proportion of formality, and the parts are in chord to the base, and not in discord to one another, but carry all the chords to fill the harmony, when they have a like tendency to their respective succeeding notes, when they imitate one another's measure, then is the consort compleat.

Only then is the ear 'pleased with the harmony' and the intellect 'amused and entertained to observe the particular parts'—'how they dance to and from the

---

[152] The implication is that contrary motion is the norm, so that parallel motion is to be avoided. Hence, in his proscription of consecutive fifths and octaves, FN differs from traditional theory, which proscribed consecutions on the grounds that they are irksome or cloying to the ear; see, e.g., Butler, *The Principles of Musik*, p. 56, and Simpson, *A Compendium*, pp. 42, 95. Descartes, *Excellent Compendium*, pp. 48–49, offered a different reason:

> these [fifths and octaves] are the most perfect [intervals], and therefore when one of them is heard, then is the Hearing therewith fully satisfied, and unlesse the attention bee presently removed from that to another Consonance, it is wholly occupied by the pleasantnesse thereof, so that it can little regard the variety, and the (in some sort) frigid Symphony of the Tune....

In his first treatment of consecutions, RN, *Cursory Notes of Musicke*, pp. 179–181, also dissents from both traditional and Cartesian explanations. Later on, he proscribes consecutive fifths, because the fifth (together with the third) defines a tonality. This also was the explanation of Huygens, *Cosmotheoros* (The Hague, 1698, tr. London, 1698), a work that RN read and admired; see f. 20v in RN's essay, 'The world', (GB:Lbl) Add MS 32546: ff. 19–24v, which is bound in a notebook with other essays; see also Wilson, *Roger North*, pp. 224–227.

key, and from one key to another, how they hunt one another, and in a manner imitate humane passions'.

'Of time, or the measures of music' (pp. 33–35)

At the very outset of this last section, North clearly states what he had previously hinted: that there is a parallel between subconscious and conscious levels of experience:

> As the music and harmony of tones depends upon the regular distance of the pulses, though the interstices are not perceived, nor can be measured by us, so when the interstices are so large that they can be measured by our sense, the observance of a regular proportion of time affords great pleasure, as in the beating of a drum or striking a pair of castanets.

Although pitch ('tone') and metre ('measure') appear to be different phenomena, because of the different way our auditory system processes them, they are alike in that the smaller periodicities—those too fast for conscious counting—are detected in sensory perception as pitches, whereas the larger periodicities—those slow enough to be consciously counted—are recognised in perceptual cognition as metres.

If music is defined as movement in time, the primary elements are the beats ('strokes'); and their rate of succession provides the tempo ('time') or unit of speed. When this unit is large, it conveys a feeling of slow movement; when it is small, it conveys the opposite. The tempo is marked off by groupings of beats into metres ('measures') that literally measure out the time by giving similar importance to beats at a given regularity. For North, the measuring out is either in duple or in triple proportion, 'the diminishing' being always in geometrical progression (p. 34). In duple proportion, for example, the metre consists of a simple oscillation between two similar but contrasted beats—a strong followed by a weak beat; and this regularity creates a feeling of accent ('emphasis'). From various combinations of metre, larger and more complex entities may be formed. But in a succession of elements (beats, 'strokes'), periodicity—rhythm—is given by the regular recurrence of accent. North thus develops Hobbes's insight that musical 'air' is diversified by 'measure' and by accent.[153]

As music moves through time, the listener's attention oscillates to and fro through the centre of oscillation—the keynote, following out a rhythmic progression that is predetermined by the music but which proceeds with

---

[153] See Chapter 1, p. 15 *supra*.

regularity, so that each stage in the process is suggested by that which precedes it and outlines a continuity that will follow and is expected. Thus (pp. 33–34):

> The due observance of time is gratefull for the same reason that I gave for the formality of a single tune, because the subsequent strokes are measured by the memory of the former, and if they do comprehend them, or are comprehended by them, it is alike pleasant. The mind cannot chuse but compare one with the other, and observe when the strokes are coincident with the memory of the former.

Whereas metre provides a feeling of stability, accent induces a feeling of change of direction. Hence, the regular recurrence of accent, as the basis of rhythm, implies a progression in cycles, in which the period of rhythm is the interval of time taken by one complete oscillation, as in a complete pendular vibration, where there are two points at which movement in one direction culminates and movement in the opposite direction begins. It is this complete oscillation that forms the unit of speed and determines the tempo.

North is silent about such notational aspects of metre as time signatures, speed directions or bar lines. Nevertheless, he presents a theoretical account of the modern time system, theoretical because modelled on the pendulum and its laws of acceleration and retardation.[154] In the modern time system, the length of a 'measure note' may vary indefinitely according to the musical tempo.[155] During the seventeenth century, when the semibreve was the measure note, there were a number of different standards for measuring its length, including the simple string pendulum advocated as a musical time keeper by Robert Hooke, Robert Boyle and others.[156] It is not without interest, therefore, that North alludes to the problem of keeping musical time, when he suggests (p. 34):

> the less the intervals are [i.e., the quicker the tempo], the more grateful is the measure, because it is easily and exactly represented by the memory, whereas a long space of time that cannot be comprehended in one thought, is not retained in the memory in its exact measure, nor can abide

---

[154] Herissone, *Music Theory*, pp. 33–34, described this section of the *Essay* as the 'earliest printed account' of the modern time system in music.

[155] Andrews, *An Introduction to the Technique of Palestrina*, pp. 31–32, provided a succinct description of the main difference between the proportional system in use from c.1400–1600 and the modern time system, which required speed directions (e.g., andante, presto, etc.). For the introduction and use of such directions in England, see Herissone, *Music Theory*, pp. 57–73.

[156] See Kassler and Oldroyd, 'Robert Hooke's Trinity College "Musick Scripts"', p. 565, and Kassler, *The Science of Music in Britain*, vol. 1, pp. 111–114. For chronometers as the standard, see Harding, *Origins of Musical Time*, and Kassler, *The Science of Music in Britain*.

> the comparison, the time past being alwayes shortned by so much as it is removed from the time present.

This problem has plagued many aspiring musicians, including North's brother, Roger, who, as a youth, desired to perform music with others—'in consort'. But he was prevented, because he had yet to acquire 'the knack of keeping time'.

> I knew the devisions and subdevisions well enough and could measure the short notes, as crotchetts etc. but long notes I could not measure, and began to thinck it impossible for me ever to attain it, for how could I by my mind make the measure of a breif [i.e., breve]?

Making his 'greif' known to Francis North, 'who was then a master', the latter showed him 'the method of doing it',

> which I took and was perfect at it imediately. It was but this; play crotchets which every one can doe, in even time by an even pass of the hand, as striking on a table or drum, then play [the same] with the ... bow, but distinguishing the notes, as in the Italians tremolo, which is easy too, in such short notes. Lastly play without such tremolous distinction, but make the distinction in the mind, which when one attempts is as easy as the rest, so by counting in the mind even measures, that may be judged so, I found I could governe the time of the longest notes. For it is impossible to guess the time of slow notes, there must be some [common] measure of them. Alone it is in the mind, but in consort it is done by the other parts; for observing others play gives you a measure for your owne.[157]

Learning thus to count musical time, Roger North was admitted into the consort.

CONCLUSION (p. 35)

Two very brief observations conclude the last, untitled section, as well as the whole of the *Essay*. Both observations concern musical *êthos*—the double power that music has of expressing and also of influencing moral character. On the one hand, there is the expression in music of 'style', which Francis North understands as idiom or dialect not only among different individuals, but also between nations and even historical periods—hence, the comparison with language. On the other hand, there are the technical characteristics of music

---

[157] RN, *Notes of Me*, p. 154.

that are treated as being significantly related to ethical terms.[158] From these two observations and from a previous hint concerning human passions (p. 32), it would seem that the aim of music is not to copy or reproduce but to imitate or represent human actions in such a way that the attentive listener gains not only pleasure but also new ethical insights.[159] Certainly, this was Roger North's approach to musical *êthos*, when he wrote:

> musick is a true pantomime, or resemblance of humanity in all its states, motions, passions and affections. And in every musicall attempt reasonably designed, humane nature is the subject, and so penetrant that thoughts, such as mankind occasionally have, and even speech it self, share in the resemblance so that an hearer shall put himself into the like condition, as if the state represented were his owne.[160]

It might seem, however, that Francis North's technical treatment of the semantics of music by means of the principle of formality has little relation to ethical terms. Yet, his second observation, although referring to the ancient Greek modes, implies that this pre-compositional principle is 'in nature'. What might this mean? If we regard the third part of the *Essay* as an extended metaphor of the mechanics of motion against a resistance, then the answer is to be found in North's conception of the auditory (ear-brain) system as itself a mechanism that works always towards equilibrium when left to its own devices. Just as the auditory system is 'in nature'—that is, within us—so too is the self-regulating principle. If this interpretation is correct, we may infer that North conceives morality not as obedience to divine or royal fiat of will but as self-governance.[161]

---

[158] See Anderson, *Ethos and Education*, pp. 1–33, 177–180.

[159] See Dryden, *Of Dramatick Poesie*, pp. 17–19, 128, 136, who from 1668 onwards published similar arguments for 'poesie'.

[160] See f. 48v in RN, 'An Essay of Musicall Ayre...', (GB:Lbl) Add MS 32536: ff.1–90 (the irony is typical RN). For an earlier version, see RN, *Cursory Notes of Musicke*, p. 150.

[161] Hobbes developed this implication in detail; see Kassler, *Inner Music*, pp. 49–107. For changing conceptions of morality during the seventeenth century, see Schneewind, *The Invention of Autonomy*. RN developed FN's model by conceiving goodness as the moral analogue of elasticity; see Kassler, *Inner Music*, pp. 160–207; and DN (2) relied on the same model to explain the balance between bullion and coin. Grassby, *The English Gentleman in Trade*, pp. 352–357, assumed that DN (2) developed his concept of equilibrium from double-entry accounting. But I would argue that such accounting prepared DN (2) to accept and then to apply FN's model to explain an economic process (money supply). For RN's use of double-entry accounting, see Parker, 'Roger North'.

3 THE CRITICS OF THE *ESSAY*

Little is known about the reception of North's *Essay* after its publication in 1677. As previously mentioned, Hooke saw sheets of the *Essay* at his bookseller's shop, but he did not purchase it, whereas Pepys, who did purchase a copy, attributed authorship to Wallis.[1] To these facts, we may add Roger North's statement that the *Essay* 'sold well, and in a few years after [1677], it was out of print, and ever since is scarce to be met with, but in private hands'.[2] And we may refer also to his brief remarks on the reception of the *Essay* in the studies of Edmund Matthews—Francis North's university tutor,[3] as well as in the publications of John Wallis and William Holder. It seems that Matthews was one who undertook the experiments called for in the epistolary introduction to the *Essay*. But all that we know of these experiments is Roger North's statement that Matthews was 'so affected' by the *Essay*, 'that he made a perpetuall comment upon it; and took much paines by experiments to explain the doctrine of pulses; but these proffers are all mislayd or lost'.[4]

In 1682 Wallis published his edition of Ptolemy's *Harmonicorum*, in which he compared the philosophies of music of the ancients and moderns and, as Roger North supposed, 'persues upon' the 'scheme' of Francis North, 'but in a different manner'.[5] If the term 'scheme' refers to the two graphs (*Figure 1.B*, Tab. i and ii), then Roger North is incorrect. But if the term refers to the coincidence theory of harmony, then his statement requires qualification, since,

---

[1] See Chapters 1, p. 28, and 2, p. 69 *supra*.

[2] RN, *Life/FN*, p. 249.

[3] See Chapter 1, p. 23 *supra*.

[4] RN, *Life/FN*, p. 250. Matthews may also have had practical interests in music, since his name appears among the subscribers to Mace, *Musick's Monument*, along with Barrow, Newton, Samuel Morland and others.

[5] RN, *Life/FN*, p. 250. See Wallis, 'Appendix', who cites no modern philosopher living after 1650; see also [Cluver], 'Claudii Ptolemæ', pp. 20-21, who describes the 'Appendix' as containing

> a brief Account of the *Ancient Harmonicks*, according to the different *Sects* of the *Authors*, compared one with another; and with the [philosophy of] *Musick* of this *Age*: Shewing how and wherein the *Greek Musick* agreed or differ'd from ours; And how the several particulars thereof, are, for the most part, retained in ours, but very differently expressed; and what in the one, answers to what in the other.

There has been no detailed study of the 'Appendix', which was reprinted in vol. 3 of Wallis, *Opera mathematica*.

there were at least four versions of that theory.[6] Judging from the method Wallis used to represent the periodicities of the unison, octave and fifth [*Figure 3.1.*], his version seems like those others that considered individual consonances without relating their frequencies to a fundamental frequency.[7]

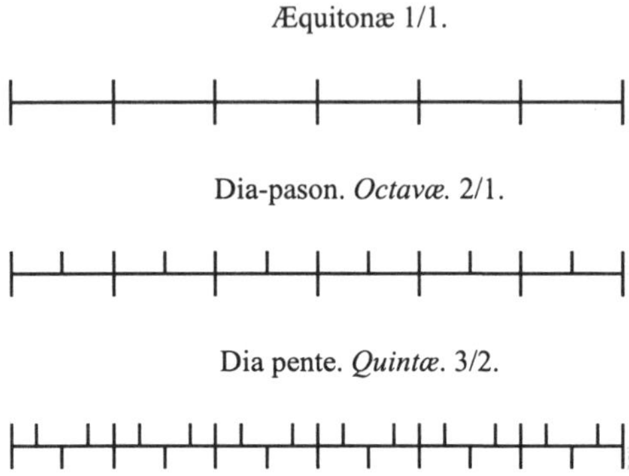

Wallis's representation of periodicities
[After Wallis, 'Appendix', p. 319]

Figure 3.1

Twelve years later, in 1694, Holder published a treatise on the 'Material' part of music—its natural philosophy.[8] Unlike Francis North, therefore, he did not 'meddle with the Artificial part'—the 'Art of Composing'—or 'the Metric and Rhythmical parts'.[9] Like North, he adopted a kinematic variant of the continuum hypothesis to explain sound propagation. He also relied on pendulum analogies to explain strings moving as double pendula, but he put no

[6] See Chapters 1, pp. 7, 9, 13 and 2, p. 52 *supra*.

[7] Wallis, 'Appendix', pp. 319–320.

[8] See Holder, *A Treatise*, [sig. B]; see also Poole, 'The Printing'. Hooke and Holder had frequent discussions about the philosophy of music, some as early as 1672; see Hooke, *The Diary*. But Holder makes no mention of Hooke, though he is indebted to some of his notions. Nevertheless, Holder presented Hooke with a copy of the treatise, and Hooke provided a synopsis of it in the Royal Society's *Transactions*. For details, see Kassler and Oldroyd, 'Robert Hooke's Trinity College "Musick Scripts"', pp. 577–579; for the reception of Holder's *Treatise*, see Kassler, *The Science of Music*, vol. 1, pp. 530–532.

[9] Holder, *A Treatise*, p. 197.

case for segmental vibrations ('breaking') or for the relative perfection of consonances by coincident pulses observed in the harmonic series. Although he recommended a variety of meantone temperament as a practical necessity, he rejected it theoretically, preferring instead to use arithmetical, not geometrical proportions, for scale division.[10] Hence, Roger North complained that the treatise is 'so puzzled with numbers, that it is uneasy to read', because Holder makes such a 'great adoe about deviding tones, major, minor, dieses, and comas, with the quantitys of them, as most authors have done, but most vanely for sounds cannot be devided, tho summes may'.[11] For Holder, therefore, scale division is determined by arithmetical proportions and not, as it was for Francis North, by judgments of sense.

Neither Wallis nor Holder cite the *Essay*, whereas the comments of three critics, shortly to be considered, take their point of departure from that work. The first two critics—one anonymous, the other Isaac Newton—wrote brief comments in 1677. These comments do not refer to North by name, so that it is possible neither critic knew the identity of the author of the *Essay*. The third critic, Roger North, began writing a longer comment in the early 1690s. Although he was well aware of the identity of the author, he, too, does not refer to the author by name, perhaps because he respected Francis North's desire for anonymity. The chief purpose of this chapter is to examine these comments not only as responses to the *Essay* but also as revealing some of the problems confronting philosophers of music from 1677 to the close of the seventeenth century.

THE 1677 CRITICS

I. NOTICE BY AN ANONYMOUS INDIVIDUAL

In the month following publication of North's *Essay*, a notice appeared in the Royal Society's *Transactions*, in which an anonymous critic, following the usual practice of that journal, provided a synopsis or descriptive summary of contents, not a proper review.[12] Nevertheless, there are two points of interest in the synopsis. On the one hand, the anonymous critic tends to distance himself from North's statements in the first part of the *Essay* (pp. 5, 6) that sound may be produced in the 'Torricellian vacuity', for when the critic refers twice to this

---

[10] Holder, *A Treatise*, pp. 196–197; for Holder's scale division, see Chenette, *Music Theory*, pp. 147–149, 169–170, 172, 383.

[11] RN, *Life/FN*, p. 250.

[12] In some cases anonymous contributors can be identified from the records of the Royal Society reproduced in Birch, *The History of the Royal Society*; unfortunately, however, no such record for this contribution seems to have been preserved.

statement, he adds: 'according to him' or 'saith he'. Even though North believed that this phenomenon is explained by his hypothesis, he initially included it in a list of other phenomena requiring experimental verification, where his statement is conditional, not absolute: sound 'may be produced in the Torricellian vacuity'. Afterwards, however, he repeated the statement in a less conditional manner: 'We see a bell will sound in the Torricellian space'. I will return to this point again when treating Newton's criticisms. On the other hand, the critic's synopsis of the second part of North's *Essay* is barely adequate, for it fails to bring out what is most original in this part. For example, he gives no indication of the improved understanding of the initiation of vibration in aerophones, especially flue pipes, nor does he mention the new definition of tone that is hinted, first, in the section on consonance and its related 'Scheme' and, subsequently, in the section on nodal phenomena—'breaking' and segmental vibrations. But he does describe the latter section as 'ingenious' without revealing why. These failures, however, might be explained by the critic's greater attention to impact as an elastic process.

As previously mentioned, in the 1650s and by 1665, respectively, Hobbes and Hooke both conceived impact on the physical model of stretched strings.[13] By 1677, however, Hooke had replaced his earlier model with a mechanical mass-spring model, conceiving bodies (not their particles) as a helical spring that is capable either of compression or extension and explaining their behaviour as a consequence of the vibratory motion of their particles.[14] As E. Williams pointed out, the helical spring illustrates the principle of elasticity in its most classic and most obvious form, for the particular material from which the spring is made is seen as something contingent:

> While it is known that the change in length of a helical spring such as Hooke used, results mainly from the torsion of the wire, the action of the spring, regarded as a unit, is nevertheless that of either direct tension or direct compression. The torsion is an intermediate element: it is the effect of the direct load and at the same time the cause of the change in length. Viewed in this way, Hooke's observation on the correlation of the primary cause and the ultimate effect holds true.[15]

---

[13] See Chapters 1, pp. 11, 17, 20–21, and 2, p. 46 *supra*; see also Kassler, *Inner Music*, pp. 2, 141–142.

[14] Hooke's new model is dynamically equivalent to a pendulum and to stretched strings. The date when he changed to this model has yet to be ascertained; but on 17 December 1677 he privately demonstrated his 'Theory of Springs and vibrations'; and in 1678 he began to write up his theory, which was approved by Wren; see Hooke, *The Diary*, pp. 334, 367–376. The result was Hooke, 'Lectures de potentia restitutiva'. For this theory and also for Hooke's recognition that elastic bodies vibrate in different ways, see Kassler, *Inner Music*, pp. 151–156, 162–165.

[15] Williams, 'Hooke's Law', p. 76.

The correlation referred to is the linear relation for elasticity known as Hooke's law—*ut tensio sic vis*: as the deflection, so the force [*Figure 3.2*].

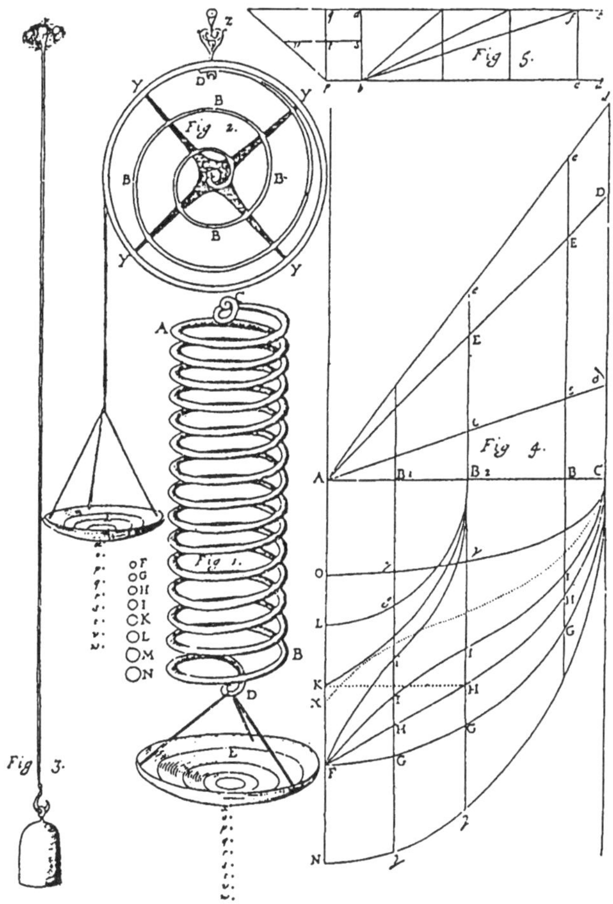

Hooke's law of elasticity demonstrated
[Hooke, 'Lectures de potentia restitutiva' (facing p. 333)]

Figure 3.2

Although bells, musical strings and columns of air in pipes are different material substances, as elastic bodies that are acoustic they all obey Hooke's law. Hence, when they are displaced from their equilibrium positions, the force of restitution—the force tending to restore them to their equilibrium positions—is proportional to the displacement, and the time of vibration is constant.

The same is true of bodies that are elastic but not acoustic, so that when the bob of a pendulum (for example) is displaced, or deflected, or pulled aside a little, the amount of the deflection is always very nearly proportional to the applied force that was used to produce the deflection. If the swings of the pendulum are not too big, the motion may be considered as simple harmonic motion, in which the accelerating force increases with the distance of the body from some fixed point or centre. Under such conditions, the swings about this centre will be made in equal times whether they are large or small ones. Although Galileo discovered these properties of the pendulum, it was Hooke who made the first attempt to understand simple harmonic motion in an elastic context.[16] Indeed, Hooke recognised that upon his law depends the correct going of clocks and watches; and he speculated that upon his law also depends the elastic vibrations giving rise to heat, light, sound and other energies. Hooke, therefore, laid the basis not only for a theory of elasticity, including the elastic vibrations giving rise to sound, but also for a new science of acoustics, according to which sound is defined as wave motion in a deformable material medium (the modern theory implies that without doubt some material medium is necessary for acoustic transmission).

II. LETTER FROM ISAAC NEWTON TO JOHN NORTH

Unlike the anonymous notice, Newton's letter of 21 April 1677 offers critical conclusions about some statements in North's *Essay*. This letter has had at least three printings: in 1869, in 1945–46 and in 1960.[17] Nevertheless, it has attracted little attention;[18] and, so far as I have been able to determine,

[16] See Truesdell, *The Rational Mechanics*, pp. 53–58. For Hooke's earlier statement of simple harmonic motion, see Chapter 2, p. 53 n.58 *supra*.

[17] See Appendix A, Editorial Note, p. 134 *infra*.

[18] Most recently, Gouk, *Music, Science and Natural Magic*, pp. 58–59, asserts, without evidence, that FN 'relies heavily on Descartes and Mersenne' and that for Newton, the *Essay* 'was an example of false Cartesian philosophy, based on a fallacious understanding of the nature of matter'.

the only examination of substance was written by C. T. Jacob in response to the second printing.[19] Because Jacob had neither seen the *Essay* nor knew the identity of its author, he limited his comments to the letter itself, proceeding on the assumption that Newton's conclusions were themselves open to criticism in light of more recent knowledge. I shall follow Jacob's examination of Newton's conclusions, before considering two further matters on which Jacob did not comment and on which Newton chose not to conclude.

At the outset of his letter, Newton drew attention to two items in North's list of the phenomena of sound in the first part of the *Essay* (p. 5). One item, the first on North's list, was the same as that singled out by the anonymous critic—that sound 'may be produced in the Torricellian vacuity'. Against this, Newton concluded that no sound would be heard if a bell were supported without touching the glass of a vacuum receiver. As Jacob pointed out, perhaps somewhat optimistically, this can be demonstrated effectively when an electric bell is sonically insulated within a receiver.[20] But he failed to point out that Newton did not offer any details about how the bell might be sonically insulated. Rather, it was Huygens who, in 1690, suggested that sonic insulation might be achieved by placing the bell on cotton or on feathers,

> in such a way that it cannot communicate its tremors either to the glass vessel which encloses it, or to the machine; a precaution which has hitherto been neglected. For then after having exhausted all the air one hears no sound from the metal, though it is struck.[21]

But this solution to the problem would be unsatisfactory, for the cotton or feathers would muffle the sound of the bell. In short, until adequate means were found for sonic insulation, there would be no way to verify either North's phenomenon or Newton's conclusion.

Newton himself may have realised this, because after 1677 he encouraged others to devise several modifications of the vacuum experiments.[22] According to Frederick Hunt, nothing significant was added to the earlier experiments by these modifications, for they merely amounted to 'replacing the vibrations of the bell by the vibrations of the wall of an air-filled receiver containing the bell'. Moreover, the vacuum experiments before, as well as after 1677, had been devised to test the effect of a vacuum on the

---

[19] See Jacob, 'Notes on Newton's Letter'.

[20] Jacob, 'Notes on Newton's Letter', p. 480; but see, e.g., Hunt, *Origins in Acoustics*, p. 115, who stated that such insulation was 'and still is!' difficult to implement.

[21] Huygens, *Treatise on Light*, p. 11.

[22] Most of these were made by Francis Hauksbee and are described in Hunt, *Origins in Acoustics*, pp. 118–120.

sounding source rather than to test whether there could be any effective interaction between the vibrating source and a highly rarefied medium. This, perhaps, was the central problem, since, as Hunt pointed out,

> an erroneous interpretation has adhered to these [vacuum] experiments with stubborn tenacity from the time of their first performance in the seventeenth century down to their present and continued use as lecture demonstrations in elementary physics classes ... [for] these experiments do *not* prove the inability of a rarefied medium to *transmit* acoustic energy, but only the extreme difficulty of *imparting* any appreciable amount of vibratory energy to such a medium.[23]

Newton also expressed doubts about another item in North's list (the fifth)—that sound 'seems to come to the ear in strait lines when the object is so scituated that it cannot come in a straight line to the ear'. As Jacob pointed out, even if Newton's doubts were true up to a point, North's observation was not faulty, because a listener still hears sound *as if* it travels in a straight line when a large streamlined object—one that is designed to reduce air or water resistance—is placed between a source and a listener.[24] Nevertheless, Newton's doubts led Jacob to infer that experiments in the reflection and refraction of sound had not yet established the principles (the same may be said of diffraction); and this inference is correct. To establish the principles, what was required was an understanding of the physical problems involved: that the 'ray' is a line along which it is imagined sound is propagated and not the reification of the path traversed by sound;[25] that rays are not curved but are always normal (that is, perpendicular) to the wave front; and that deviations from rectilinear motion are a property of a motion propagated in an elastic medium and are not caused by the motion of air or winds. Without recognition of the relation between the concepts of the mathematical (ideal) ray and the physical (real) wave front, little progress could be made in the theory of the propagation of sound by waves.[26]

In 1672 Newton seems to have been thinking about the problem of wave fronts and the direction of their propagation, when he wrote in a private letter:

> to me the fundamentall supposition it selfe seems impossible; namely that the waves or vibrations of any fluid can like the rays of Light be

[23] Hunt, *Origins in Acoustics*, pp. 120–121.

[24] Jacob, 'Notes on Newton's Letter', p. 480.

[25] A ray is now defined as a set consisting of point *P* on a straight line and all points of the line on one side of *P*; hence, it is a mathematical concept.

[26] For some of the problems, see Shapiro, 'Kinematic Optics', especially p. 248; see also Shapiro, 'Light, Pressure and Rectilinear Propagation', pp. 286–289.

> propagated in streight lines, without a continuall & very extravagant spreading & bending every way into ye quiescent Medium where they are terminated by it. I am mistaken if there be not both *Experiment & Demonstration* to the contrary.[27]

It is possible, therefore, that North's *Essay* restimulated his thinking on this problem, since nine years later he wrote: '*Every tremulous body in an elastic medium propagates the motion of the pulses on every side straight forwards.*'[28] According to Alan Shapiro, this 'enigmatic proposition' makes the important point that in a uniform elastic medium the rays are normal (perpendicular) to the wave front.[29]

The rest of Newton's letter concerns the second part of the *Essay* and, in particular, the statement (p. 8) that equal vibrations of two strings work one another to a synchronism, the rationale for North's coincidence theory of harmony. But Newton concludes: 'it seems otherwise to me'; and in support of this conclusion, he begins with the manner in which sound is transmitted. According to him,

> the pulses must not be supposed to extend each of them at once from the sounding body to the ear in such manner that all the air in that interval be moved together first forward and then backward, and so forward and backward again so long as the sound lasts:

Rather,

> the pulses are to be conceived like so many spherical concentrick waves whose center is the sounding body and which arising continually from that center dilate and flow on from thence with that swiftness we find sound propagated till they arrive at the ear; new pulses from the center continually succeeding them after the manner of undulations made by throwing a stone into water.

Jacob supposed that the first statement above quoted refers to the action of two equal strings, which move the air 'forward and backward again', a belief undermined in the twentieth century by aerodynamic experiments.[30] But, if this is Newton's meaning, then Hooke and North both grasped that the velocity of a vibrating string actually is quite slow and that at such a low

---

[27] Newton to Oldenburg for Hooke, 11 June 1672; quoted in Shapiro, 'Light, Pressure and Rectilinear Propagation', p. 287.

[28] Newton, *Mathematical Principles*, p. 371.

[29] Shapiro, 'Light, Pressure and Rectilinear Propagation', p. 288.

[30] Jacob, 'Notes on Newton's Letter', pp. 480–481.

velocity the kinetic energy of a string is infinitesimal in comparison with the acoustic output from the soundboard or resonator. Hooke, for example, described the phenomenon that a string vibrates silently unless mounted on an effective resonator or soundboard; he then concluded that a string causes 'strokes' or shocks in the soundboard and suggested that these are propagated through the air as a pulse front.[31] Indeed, Jacob himself cited Hooke as one seventeenth-century investigator who sought to refute the widespread belief that a tone was produced directly from the string.

It seems to me that Jacob misunderstood Newton's point, perhaps because he did not have access to the relevant statements in the *Essay*, for North did not claim that all the air is moved 'first forward and then backward, and so forward and backward again'. What he did claim is that for tone, an elastic impact causes the air particles to oscillate like a swinging pendulum, so that the state transmitted is a pulse front of constant phase. Against this, Newton proposes that the pulses be conceived as intersecting spherical surfaces and, hence, as physical wave fronts. In so doing, he opens up the problem of relative timing—phase relations of the sound waves—in order to criticise the coincidence theory. This theory, in all its versions, assumed that two strings begin vibrating at the same instant, so that their pulses will coincide precisely at regular time intervals.[32] The problematic nature of this assumption was recognised as early as 1654, when Charleton surveyed a number of answers to the question: '*in what instant an Harmonical Sound, created by a Chord of an instrument percussed, or abduced from its directness, is begun?*'.[33] Although North specifically referred to a situation where two strings do not begin vibrating at the same instant, he supposed that, because of resonance, the strings will work one another into synchronism after the second or third pulse, so that the listener is affected equally and in constant phase by the two sound sources.[34]

To criticise this in-phase supposition, Newton's proposal of intersecting waves as the appropriate analogue for sound propagation leads him to a premiss and two conclusions as follows:

> the pulses of two sounds may at the same time strike one ear both together and another ear alternately and yet do exhibit the same concord

---

[31] See Chapter 2, pp. 70–71 *supra*.

[32] See Cohen, *Quantifying Music*, p. 232, who focused on this aspect of the problem. But this is to ignore other aspects. For example, Galileo assumed that if two sounds are 'commensurable'—have a common measure between their vibrations—they automatically are heard in phase; see Galilei, *Two New Sciences*, pp. 99, 101.

[33] Charleton, *Physiologia*, pp. 228–229.

[34] Chenette, *Music Theory*, p. 186.

> to all ears, it follows 1st that concords arise not from the coincidence of pulses at the ear nor have any dependance on such coincidences, and 2dly that unisons are rather a harmony of two like tones then a single tone made more loud and full by the addition as the author would have it, p. 8, l. 20.

Note that Newton's premise has to do with auditory localisation and the problem of binaural hearing: can the two ears of a listener discriminate the relative positions of tonal stimuli and, if so, how? This problem was solved only very recently, so that even in 1884 Lord Rayleigh (J. W. Strutt) lamented: 'Every day we are in the habit of recognising, without much difficulty, the quarter from which a sound proceeds, but by what step we attain that end has not yet been satisfactorily explained.'[35] By the 1920s it was known that the difference of phase between the two ears is one of the bases of auditory localisation; but this conclusion was suppressed because of certain presuppositions that guided experimental psychology during the nineteenth century.[36] Indeed, Jacob himself seemed unaware of this conclusion, since he supposed that the problem of auditory localisation was psychological—the 'mind' does not merely submit to the mechanical reactions in the ear but possesses the power of selecting from simultaneous aural sensations.[37] Hence, he concluded that North's observation 'may not have been at fault'.

We now know that listeners perceive sound as arising in three-dimensional space, not just in a flat horizontal plane; and they can distinguish if a sound is above, below, in front, to the side, or behind. But how is this done with only two ears? The answer is that the auditory (ear-brain) system can determine the direction of a sound source, because it utilises changes in the frequency distribution of a complex sound as it arrives at the ear canal.[38] This distribution changes as a result of the outer ear, or pinna, which acts as a baffle and resonator specifically to create the effect. The brain then interprets these subtle variations in frequency distribution as directional clues, rather than as changes in the timbre of the frequency of the sound source. This permits the auditory system to resolve the direction of a sound source in three-dimensional space. Without this capability listeners would be unable to distinguish between sounds coming from the front and rear.

In spite of a lack of knowledge about auditory localisation, Newton's premiss led him to conclude against the coincidence theory of harmony and against North's statement (*Essay* p. 8):

---

[35] See Alexander Ellis in Helmholtz, *On the Sensations of Tone*, p. 535.

[36] For a brief review, see Boring, *Sensation and Perception*, pp. 381–392.

[37] Jacob, 'Notes on Newton's Letter', p. 481.

[38] See Kraemer, 'Two Speakers', pp. 71–74.

> Where the terms are equal [i.e., 1 to 1] and the coincidence is at every pulse, it may be said [to be] the same sound though proceeding from several instruments. For the mixed sound has no alteration, but by being more loud and full by the addition.

The rationale for this statement is North's mechanical principle of resonance: pulses that synchronise frequently—and which are comprehended in the same time—reinforce the sound heard. This principle explains his statement about unisons (above), as well as his statement about octaves (*Essay* p. 10), where the sound is augmented 'by the addition of stronger pulses at proportionable intervals'. It also explains his statements about coupled systems (*Essay* pp. 16, 20), where the sound of an acoustic vibrator is reinforced by the presence of a resonating surface; and about music in parts (*Essay* p. 24), where the 'base' is louder than the 'other' accords. And it also provides the rationale for North's statement (*Essay* p. 8) that equal vibrations of two strings work one another to a coincidence or synchronism. We might infer, therefore, that Newton's chief purpose is to undermine North's mechanical principle. If this was his intention, then his comments have wider import, since resonance is the main mechanism of propagation in kinematic variants of the continuum hypothesis, a tradition that was introduced into England by Hobbes.[39]

Jacob was silent about North's mechanical principle as well as about Newton's two other remarks on the *Essay*. In one remark, Newton supposes—referring to the *Essay* p. 12—that the 'explication' of sound production in 'whistles' is 'ingenious' but 'not altogether substantial'. But Newton does not offer any conclusions of his own, because of a lack of experiments. Since North does not specifically mention 'whistles' until page 14 and since he treats 'the cause of the tone of a pipe, and the pulse that gives the sound' from page 12 onwards, it seems that Newton's term 'whistles' denotes aerophones in general, not merely flue pipes. It is not without interest, therefore, that a decade later he explicitly stated the correlation between wavelength and pipe length.[40] In the other remark, Newton describes North's treatment of the 'breaking of tones into higher notes' (*Essay*, pp. 17–21) as 'ingenious' and 'judicious'—high praise. But, once again, he does not offer any conclusions of his own, because, as he wrote, 'I want experience to discern whether altogether solid'. As previously indicated,[41] since North's treatment of 'breaking' provides an explanation of overblown and flageolet tones, as well as harmonics, his *Essay*

[39] See Hesse, *Forces and Fields*, p. 118; see also Kassler, *Music, Science, Philosophy*, p. 121. For the understanding of resonance before 1677, see Green, *The Harmonic Series*, pp. 98–151.

[40] Newton, *Mathematical Principles*, p. 384, (in the third edition referring to experiments by Sauveur on the velocity of sound) concludes: 'it is probable that the breadths of the pulses [i.e., wavelengths], in all sounds made in open pipes, are equal to twice the length of the pipes.'

[41] See Chapter 2, pp. 62–68 *supra*.

marks the beginning of a new definition of tone as a composition of simple pendular vibrations, for he recognised that an elastic body can vibrate in more than one mode at once. Moreover, two months before Wallis reported on the discovery of nodal phenomena, North (*Essay* pp. 17–19) associated harmonics and nodes in the context of a discussion of the trumpet marine, where he also provided a brief comparison between that instrument and the natural, valveless trumpet.

This same comparison formed the basis of a paper that scholars have cited almost as frequently as the report of Wallis. Published in 1692, its author, Francis Roberts,[42] not only clarified how the air in the trumpet breaks up into separate vibrations by reason of the similarity of the trumpet's notes to those of the trumpet marine but also sought to explain why the trumpet produces only a discrete set of pitches and why some of these pitches are 'imperfect'.[43] He used pitches of a scale in just intonation to show that the trumpet and trumpet marine, each having natural frequencies in the ratios 1, 2, 3, 4, ..., are thus incapable of other 'musical' vibration. He also pointed out that if the trumpet marine is stopped at any point not making one of the two resulting portions of the string an aliquot part of the other, 'the vibrations of the parts will cross one another, and make a sound sutable to their motion, altogether confus'd'.[44] Roberts mentioned the experiments involving resonance between strings that were not in unison and correctly described the nodal forms of a vibrating string observed by means of a paper rider as described by Wallis (though Wallis is not named).[45] He concluded that the requisite for a string to sound a 'synchronous' tone is that it must sound the pitch of an aliquot part of the string. When this happens, '(the vibrations of the equal parts of a string being *Synchronous*) there is no contrariety in their motion to hinder each other'.[46]

---

[42] Francis Roberts [Robartes], a virtuoso and amateur musician, was a son of John, viscount Bodmin and Earl of Radnor, who, with FN, was appointed to the reorganised Privy Council (21 April 1679) and served as lord president (1679–84). FN and RN may have known Roberts either through this connection or through the Gentleman's Society, a group of amateur and professional musicians who from c.1674 met weekly or oftener to play music in the French style of Jean-Baptiste Lully; see *NP 4*, pp. 6, 14. See also RN, *The Musicall Grammarian*, p. 265, according to whom, pp. 261–262, 'all the compositions of the towne were strained to imitate' Lully's style, and 'none came so neer it as the honourable and worthy vertuoso, Mr. Francis Roberts'. From 1672–73 Roberts was a member of parliament for various Cornish boroughs or for Cornwall; from 11 December 1673 he was a fellow of the Royal Society; and from c.1705 he was one of the tellers of the exchequer. Hooke, who met Roberts on several occasions during the 1670s, describes him as 'very ingenious'; see Hooke, *The Diary*, p. 83.

[43] See Roberts, 'A Discourse'; see also Chenette, *Music Theory*, pp. 44–45, 80–81.

[44] Roberts, 'A Discourse', p. 561.

[45] See Dostrovsky, 'Vibration Theory', pp. 205–206, and Truesdell, *The Rational Mechanics*, pp. 120–121, 294.

[46] Roberts, 'A Discourse', p. 561.

Like Wallis, however, he seems to have assumed that string length is the only important factor that controls timing and determines pitch.

Not long after Roberts' paper was published, Roger North made a copy, interpolated some corrections to the text, and then, because Roberts' norms were mathematically determined, added two 'Memorandums as to the phisicall solution of this theory'.[47] It seems, however, that he did not come across Newton's letter until sometime in the mid-1720s.[48] But he immediately recognised that Newton's criticisms tended 'seemingly to dissolve our whole hypothesis of harmony, and coming from such an author, will not easily be diluted but requir[e]s a carefull answer to be made to it'.[49] In his 1726 draft to answer the criticisms, he began by summarising that part of the letter in which Newton raises the problem of phase relations.[50] He then proceeded with his answer by way of six 'notes', the substance of which is as follows:

(1) that 'all unison tones however the bodys are scituate with respect to each other, puls equally and at the same moments of time, for the [vibratory] force of each other upon the respective springs reduce all to that state';
(2) that 'whither the pulses strike the ear *eodem instante* or not, the memory joynes sensations and thereby the mind judges the consonance alike; for it is the syncronism of each cours, and not the exact time of the touch, that makes the consonance';
(3) that 'the mind judgeth of place from whence sound comes, by the manner of its centration, for if obstacles break the cours of that, the judgment of place is fals';
(4) that 'the objector is in an error saying that a single tone is not made more loud by the addition of an unison';

[47] For his copy of Roberts' paper and the added 'Memorandums', see (GB:Lbl) Add MS 32549: ff. 31–35, which is described with an analytical index in *NP 2*, pp. 28–35. Roberts' paper formed the basis of RN's later treatment of the modes of vibration in the 'tube trumpet'. For his developing ideas, see (in chronological order) *NP 2*, p. 194, *NP 5*, pp. 44, 161–162; ff. 26, 51–56, 67v, 69v, 70–74v, 79 in 'The Theory of Sounds...' (GB:Lbl) Add MS 32534: ff. 1–82v; *NP 3*, pp. 164–166; see also Kassler, *Inner Music*, pp. 165–166.

[48] RN refers to Newton's letter in three of his writings: (1) 'Theory of Sounds...1726', (GB:Lbl) Add MS 32535: ff. 1–73v; (2) 'Theory of Sounds...1728', (GB:Lbl) Add MS 32535: ff. 74–149; and (3) *Life/FN*, p. 250. For a description of the two manuscripts, with a combined analytical index, see *NP 3*, pp. 48–178.

[49] See f. 118 in RN, 'Theory of Sounds...1728', (GB:Lbl) Add MS 32535: ff. 74–149.

[50] ff. 47–48v in RN, 'Theory of Sounds...1726', (GB:Lbl) Add MS 32535: ff. 1–73v. In concluding RN offers a few reflections 'on our manner of judging many impressions from sence, wherein wee are not a litle wildred by mistaking meer memory[s] for actuall sensation.'

(5) that, according to Newton, 'unisons are rather an harmony of like tones, by which he seems (after his way,) to decline the mechanick, for he says not how, or in what respect, like; and to set up [occult] quallitys in harmony';
(6) that 'the consonances or dissonances of sounds, reside not in the bodys that emitt the tones, or in the order of pulses with respect to each other; for harmony is an idea of an interior result of the mind'.

In the first 'note', North restates his brother's conception of mechanics as a study of motion against a resistance, whereby motions, when opposed, will order themselves so as to balance or equalise the distribution of force.[51] In the second and sixth 'notes', he offers his own solution to the problem of phase relations, insofar as this problem relates to the sensory perception of consonances. Since this problem could not be solved in the seventeenth century or even for some considerable time afterwards, it is worth pointing out that in 1784 Matthew Young provided an explanation similar to North's, as follows:

> It has been very acutely objected to the doctrine of coincidences, that if the first pulses propagated from the sounding bodies do not arrive at the ear at the same instant, they will in some cases never coincide. This we must acknowledge is true; but then there will in such cases be always two pulses which will very nearly approach to an actual coincidence, and as the effects of all impressions on the organs of sense are not instantaneous, but continue for some time, these pulses, when they make their nearest approach to each other, will have the same effect as if they had actually coincided: and after they have made this their nearest possible approach, they will continue afterwards to preserve that interval, because the two bodies perform their proper vibrations in the same time, and consequently the same impression will return as regularly as if they had actually coincided.[52]

In the fourth 'note', Roger North reasserts the resonance effect of strings vibrating in unison; and in the fifth 'note' he refers to the second of the two conclusions that Newton drew from his premise about auditory localisation.[53] He complains that Newton's statement sets up 'quallitys', perhaps because North had studied the second, 1713, edition of the *Principia* with the preface by Roger Cotes, who reinforced the notion that gravitational attraction was to

---

[51] See Chapter 2, p. 46 *supra*.

[52] Young, *An Enquiry*, pp. 199–200. Maley, *The Theory of Beats*, pp. 22–23, attributed the problem of phase relations to an oversimplified model of sound as the propagation of pulse fronts, perhaps just as Newton did. But the problem, historically speaking, seems more complex.

[53] See this Chapter, pp. 98–99 *supra*.

be regarded as an action at a distance and, hence, as an occult rather than a manifest quality capable of mechanistic explanation.[54] Whatever the reason, Newton's second conclusion—'that unisons are rather a harmony of two like tones then a single tone made more loud and full by the addition'—lacked clarity. But his meaning might be extracted from the new theory of light and colour that he made public from 1672 onwards.

According to that theory, colours exhibited by the spectrum are simple—that is, have each a unique frequency, whereas sunlight (white light) is a compound of all colours.[55] In the early statements of this theory, Newton several times referred to colour 'harmony'.[56] In 1675 he speculated that colours form 'harmonies' just as tones do. To elucidate his meaning, he divided the spectrum into seven segments that were proportional to a string sounding the seven tones of the diatonic scale in just intonation.[57] And he repeated this analogy in the various editions of his *Optics*, published first in 1704.[58] It would seem, therefore, that Newton's terms 'harmony' and 'tones' denote, respectively, arithmetical proportions and frequencies. Hence, in the case of two strings sounding in unison, the 'tones' are 'like', because their frequencies are in the same proportion, 1:1. For Francis North, however, two sounds in unison are, physically, 'the same sound' reinforced by resonance (his reason why unisons do not function in music as concordant intervals).[59]

THE LATER CRITIC

When, in the 1690s, Roger North began writing 'Some Notes upon an Essay of Musick', he had passed through a period of considerable personal distress, during which he lost his closest friends—his brothers John, Francis and Dudley—as well as his public appointments. Some of his distress is hinted in

---

[54] For Cotes's preface, see, e.g., Buchdahl, 'Gravity and Intelligibility'; for RN and Newton, see *NP 6*, pp. 11–16.

[55] Shapiro, 'Kinematic Optics', p. 207, and 'The Gradual Acceptance of Newton's Theory', p. 65.

[56] See Shapiro, 'Artists' Colors and Newton's Colors', pp. 610 n. 25, 619.

[57] See Newton, *Letters and Papers*, pp. 192–193. For some problems resulting from this analogy, see Shapiro, 'Artists' Colors and Newton's Colors', p. 626.

[58] Newton, *Opticks*, pp. 125–129, 154–158, 210–212.

[59] Newton's analogy between the spectrum and the diatonic scale in just intonation has Pythagorean implications, so that he, too, may not have considered the unison as a consonance. But his reasons would be mathematical, not physical, namely, that one—unity—is not a number but the basis of numbers. For a more overt instance of his Pythagoreanism, see Newton, *The Correspondence*, vol. 4, pp. 274–275.

the defence of suicide, which he included in his autobiographical *Notes of Me*.[60] But there, as well as in 'Some Notes', North intimates that it is better to be than not to be. Hence, both works are life affirming and point towards new directions and new projects. One project, mentioned in the former work, concerned 'matters' in his brother's *Essay*, about which he intended 'to inlarg a little upon ... in some notes upon that treatise, if it be printed againe, so pass 'em by here'.[61] Since he was then involved in editing, and writing an important preface to, a treatise by another brother,[62] it is possible that he also planned to reprint the *Essay* in a similar format. Whatever the plan, Roger North soon realised that the hints in the *Essay* required full development, not merely critical comment and amendment; and thus began a series of remarkable writings on the philosophy of music, to which he was still adding and modifying close to the time of his death in 1734.

Between 1677, when the *Essay* was published, and the 1690s, when North began writing his comment, Hooke had published his 1678 law of elasticity; and there also were a number of other advances in the science of sound. Two advances deserve mention here as background to North's comment. One advance, which began with the publication of the first 1686 edition of Newton's *Principia*, concerned the dynamics of wave propagation, an understanding of which depends on the solution to the equation of motion (F=ma).[63] Although Newton had rejected a conjectural in favour of a 'mathematical way' of philosophising, in Book II, Section VIII, he introduced a continuum hypothesis to compare the propagation of sound with pulses produced when a vibrating body moves the adjacent portions of the surrounding medium and these in turn move those next adjacent and so on.[64] He went on to make some arbitrary assumptions, the principal one being that, when a pulse is propagated through a fluid, the particles of the fluid always move in simple harmonic (in-phase) motion—or, as he put it, are 'always accelerated or retarded according to the law of the oscillating pendulum'.[65] And he proved the theory that if this is true for one particle, it must be true for

---

[60] This is the most remarkable example of RN's writing as an act of remembrance, although almost all his writings may be so regarded, including those on the philosophy of music.

[61] RN, *Notes of Me*, p. 160.

[62] This was the *Discourses upon Trade* by DN (2). The colophon is dated 1691, but the book could have been printed as late as 25 March 1692; see Letwin, *The Origins of Scientific Economics*, p. 266. For a reprint of RN's preface, see Grassby, *The English Gentleman in Trade*, pp. 352–357.

[63] The solution was found in the post-Newtonian period of rational mechanics; see, e.g., Lindsay, 'The Story of Acoustics', pp. 636–637.

[64] Newton, *Mathematical Principles*, pp. 367–384.

[65] Newton, *Mathematical Principles*, p. 375.

all the adjacent ones.[66] North, however, seems not to have studied this part of Newton's work until after publication of the second edition in 1713.[67]

The second advance took place in the understanding of the kinematics of wave propagation, which is independent of the dynamics. This understanding reached its culmination in 1690 with the work of Huygens, who formulated the concept of a wave front (common tangent) in its 'most complete mathematical and physical generality'—that sound and light spread in air by a succession of spherical surfaces.[68] Huygens specifically stated that he used the term 'waves' for the movements of sound and light

> from their resemblance to those which are seen to be formed in water when a stone is thrown into it, and which present a successive spreading in circles, though these arise from another cause, and are only in a flat surface.[69]

He also explained how two sound or light rays (lines perpendicular to the tangent) could travel in opposite directions in the same space or cross at an angle without hindering each other;[70] and this explanation led to his famous principle of secondary wavelets, according to which each particle in the path of a disturbance is a centre from which the disturbance spreads outwards through all the particles in contact.

Although his formulation of the wave front occurs in a treatise on light, Huygens made frequent reference to, or comparisons with, sound. For example, he noted that although both sound and light are propagated in time by compressional or longitudinal waves,[71] they differ in three principal properties. First, they are differently generated: sound, by the motion of the entire source, and light, by the motion of every point of the source. Second, their propagation takes place in different media: sound, through the air, and light, through the æther that freely penetrates all matter. Third, their mode of propagation differs: sound is propagated through a medium consisting of particles that are not in

---

[66] For the proofs and some of the problems, see Hunt, *Origins in Acoustics*, pp. 149–150.

[67] See *NP 6*, pp. 3–39.

[68] Shapiro, 'Kinematic Optics', p. 137, referring to Huygens, *Treatise on Light*, the first part of which was completed by 1679; the second part, written later, contains some remarks on Newton's work (but this part is not included in the English translation).

[69] Huygens, *Treatise on Light*, p. 4. For water, the 'cause' is either surface tension for small wavelengths or gravity for larger wavelengths.

[70] Huygens, *Treatise on Light*, p. 17. For a similar statement, made in the 1680s, see Hooke, *The Posthumous Works*, p. 133.

[71] Light waves are transverse, not longitudinal; but this was not understood until the early part of the nineteenth century.

contact except during collision, whereas light is propagated through a medium consisting of particles in contact.[72] Huygens also assumed that the particles of both æther and air are 'perfectly hard'. But since perfect hardness, for Huygens, meant perfectly elastic, he had recourse to Hooke's 1678 law (without naming Hooke), according to which bodies that 'act as springs', 'whether compressed little or much', recoil 'in equal times'.[73] Hence, velocity of propagation is a function of the elasticity of the medium and is independent of the strength of the compressing force.

It has been supposed that Huygens concluded the tradition of a kinematic variant of the continuum hypothesis, a tradition that began in England with Hobbes.[74] This supposition may be true when applied to *published* writings on optics and acoustics, so that Huygens' treatise did sink into oblivion during the eighteenth century. But the treatise seems to have made a deep impression on Roger North, who took a more qualitative approach to the kinematics of wave propagation not only in 'Some Notes' but also in later unpublished writings on the philosophy of music.[75] Just when North read Huygens' treatise is uncertain; but in 'Some Notes' there are improvements to the first part of his brother's *Essay* that might derive either from Hooke or from Huygens, since their kinematic variant of the continuum hypothesis was developed concurrently and reveals both similarities and differences.[76] But Huygens' treatise had one advantage, for it commenced with a succinct summary of a conjectural or probabilistic way of philosophising, as follows:

> whereas the Geometers [like Descartes] prove their Propositions by fixed and incontestable Principles, here the Principles are verified by the conclusions to be drawn from them; the nature of these things not allowing of this being done otherwise. It is always possible to attain

[72] Huygens, *Treatise on Light*, pp. 4, 7, 10–12.

[73] Huygens, *Treatise on Light*, p. 27; see also pp. 13–15, *et passim*.

[74] Shapiro, 'Kinematic Optics', p. 252 and n.381.

[75] RN never cites Huygens by name in his writings on the philosophy of music, even though as early c.1698–c.1703 he adapted to sound Huygens' explanation of how light rays can travel in opposite directions in the same space or cross at an angle without hindering each other; see RN, *Cursory Notes of Musicke*, p. 57. He thus noticed one of the most striking features of wave motion, namely, that two waves can apparently pass through each other without being destroyed; indeed, they emerge apparently unchanged. This feature is a particular case of the principle of superposition, which states that the result of combining two physical quantities, such as the amplitudes of two vibrations, can be obtained by adding the quantities as though they remain independent. The principle also implies that the combined effect can be re-separated into the two original effects.

[76] Hooke's range was wider, Huygens' focus more exact. A detailed comparison of their philosophies, not just their optical theories, remains a desideratum.

> thereby to a degree of probability which very often is scarcely less than complete proof.[77]

This way of philosophising would have appealed to North, not only because it was his brother's way but also because it suited North himself, who, by profession, was a lawyer.[78] Moreover, it also suited North's constructive scepticism, since a number of his writings reveal not only that he distrusted the power of reason to attain truth but also believed that complete certainty could never be achieved in the study of nature.[79]

Keeping this context in mind, we may turn to North's comment. To begin with, the text is undated; and the paper on which it is written offers no evidence for a date.[80] But two kinds of evidence enable us to determine a possible date. First, there is the familiar style in which the comment is written, for North introduces at least one aspect of Norfolk dialect—''em'—an abbreviated form of 'them'. As he explained in *Notes of Me*, in Norfolk

> the word (them) is not sounded, as in the north nor west; but midling; for they give a vowel more then the west, and say (heard 'em), (gave 'em), (beat 'em). And in stead of the 'n put em. In Norfolk for notice, they will say notedg. And allways instead of ready they say fitt, as fitt to fall downe, mony fitt to be payd. There are many other varietys of words, and manner of pronounciation ... but these are the cheif....[81]

North's use of the Norfolk 'em suggests that he wrote his comment after purchase of his estate in Norfolk and, most likely, after the work of remodelling had commenced in 1692. Second, four of his manuscripts, written after he settled in Norfolk, are related to his comment. These manuscripts are as follows: (1) 'Notes of Me', written between c.1693 and c.1698;[82] (2) 'Of Building' (GB:Lbl) Add MS 32540: ff. 1–68v, dated c.1695–96; (3) 'Cursory

---

[77] Huygens, *Treatise on Light*, p. vi.

[78] Huygens, *Treatise on Light*, pp. vi–vii.

[79] See, e.g., RN, 'Reason', (GB:Lbl) Add MSS 32526: ff. 120–204, transcribed in Korsten, *Roger North*, pp. 198–204. Although dated 1732, this essay seems to echo Matthew Hale, who, according to Sampson, 'Laxity and Liberty', pp. 86–87, defended legal reasoning and the common law as the accumulation of past probabilistic wisdom, arguing that, since the subject of laws are moral actions, there cannot be expected precise certainty of evidence, as in mathematics. Nevertheless, he believed that it was necessary to have 'a certaine and determinate Law' by which men could regulate their lives and that this law was given its perfect expression in the Gospels (i.e., the law of charity or 'golden rule').

[80] For a description of the paper and binding, see *NP 2*, p. 2, and *NP 4*, p. 57.

[81] RN, *Notes of Me*, p. 191. For some background to RN's interest in 'dialects', see *NP 6*, pp. 57–62.

[82] See Peter Millard in RN, *Notes of Me*, pp. 57–60.

Notes of Building', dated c.1698;[83] and (4) 'Cursory Notes of Musicke', dated c.1698–c.1703.[84] Since, in *Notes of Me*, North merely indicated that he might write 'some notes' on his brother's *Essay*, we may assume he commenced writing after c.1693 and before c.1698. But the period can be narrowed to about 1694, since the text, 'Of Building', is a draft for 'Cursory Notes of Building', which is contained in the same notebook as 'Cursory Notes of Musicke'.[85] Moreover, all three texts contain more 'finished' versions of the hypothesis that North outlines in 'Some Notes' with reference to both music and 'building'.

In his comment North includes a brief but illuminating character of Francis North (f. 44) that stresses his singularity in possessing both procedural and philosophical knowledge of music. Hence, as Roger North states, his brother differed from 'scollars' of music who 'wanted acquaintance with the subject itself', as well as from musicians and 'mechanicks' who were 'litle addicted to natural inquisitions, or experiments'.[86] Most of the comment consists of three notes that are restricted to the first and second parts of the *Essay* concerning 'matters' in (1) the section on the hypothesis of sound and some of the related phenomena; (2) the sections on tone and consonance; and (3) the sections that treat aerophones. But North also offers amendments to, as well as improvements on, some of his brother's explanations. In addition, he seeks to clarify his brother's meaning in certain instances. And he also outlines an hypothesis that, subsequently, will form a guiding principle of his own philosophy of music, a philosophy that he developed in a series of drafts, essays and treatises over a period of nearly forty years.

I. THE NOTES

In his first note (f. 42), North begins by considering his brother's Phenomena 7 and 9 (*Essay* p. 5) on the propagation of sound with and against the wind. To represent a sound wave travelling with or against the wind, he has recourse to the analogy of circular water waves, even though he conceives dilating sound waves not as circles but as spherical surfaces. Like the early investigators,

---

[83] RN, 'Cursory Notes of Building', *Of Building*, pp. 1–103; see Howard Colvin and John Newman in RN, *Of Building*, pp. xxiii–xxv.

[84] RN, *Cursory Notes of Musicke*, pp. [62]–[65]; see also *NP 2*, pp. 86–87.

[85] See *NP 2*, pp. 85–86.

[86] RN himself cultivated both kinds of knowledge; see Kassler, 'Roger North', pp. 54–55. See also the 'Premise' to RN, *Cursory Notes of Musicke*, p. iii, where he wrote: 'as I ever bent my thoughts with more than ordinary intentness to the search of truth in the way of phisicks, it could not be expected, but I should take in my darling [music], as a subject to work upon'.

including his brother, North ignores environmental conditions such as temperature, humidity and wind velocity.[87] We now know that the layers of air in contact with the ground will be retarded by friction and so will move less quickly than higher layers. Hence, if the sound waves are travelling with the wind, the upper edges will move more rapidly, whereas if the waves are travelling against the wind, the reverse is true; and this accounts for the fact that sound carries much better with than against the wind.[88]

North then considers in more detail his brother's mixture theory of the atmosphere, which Francis North conceived as consisting of three kinds of matter of increasing smallness. The gross particles—air *simpliciter*—are elastic; whereas the middle- and small-sized particles are inelastic, the one being the medium of sound, the other of light. Against this mixture theory, Roger North attributes elasticity to *all* parts of the air—the æther as well as other parts that are 'perfectly sphericall, if not the most minute'. This suggests indebtedness to Huygens' mixture theory, according to which all particles are perfectly elastic and consist of (*a*) components of ordinary bodies and air—the medium of sound, (*b*) æther particles—the medium for light, (*c*) even more subtle matter—the medium and cause of gravity, as well as of magnetic and (possibly) electric phenomena.[89]

By making all aerial matter elastic, North supposes he has facilitated the solution to the problem of whether sound can be produced in the Torricellian or 'baro[s]cope-vacuity'—Phenomenon 1 of the *Essay*. But this conclusion was to remain a matter of contention, for, even though the experimental tests of Boyle and others had failed to conclude one way or the other, Huygens accepted that sound would be interdicted in the vacuum experiments if the bell in the glass receiver were sufficiently insulated. For him, therefore, aerial particles—the medium of sound—could not penetrate glass. But the situation is otherwise for light, since Huygens cited 'the experiment of Torricelli' as a proof that ætherial matter—the medium of light—'not only passes through solids, but does so even with great facility'.[90] This provides a context for North's assertion that 'to some elastick matter, glass is pervious, and to other, not'.

---

[87] William Derham, in a paper published in 1708, may have been the first Englishman to consider environmental conditions; see Lindsay, 'The Story of Acoustics', p. 635, and Hunt, *Origins in Acoustics*, pp. 111–112. For RN's citations of Derham's work, see (in chronological order) ff. 26v–27 in 'The Theory of Sounds...', (GB:Lbl) Add MS 32534: ff. 1–62v; and *NP 3*, p. 61.

[88] See Wood, *The Physical Basis of Music*, pp. 8–11.

[89] Huygens, *Treatise on Light*, pp. 14–15, differentiated the (unobservable) particles in each group not by shape but by size (increasing smallness) and velocity, whereas RN (f. 42v) refers not only to 'spherical' but also to 'angular' particles.

[90] Huygens, *Treatise on Light*, pp. 29–30.

What, then, is the criterion of elasticity? As mentioned previously, Francis North's criterion is separation of aerial particles on impact.[91] But Huygens' criterion is the effort of aerial particles to escape from the place where, on impact, they are squeezed together at the regions of compression along the propagating waves. In spite of these different criteria, both hold that sound is produced by restitution upon compression. For Roger North, however, the criterion is the turning motion of angular particles, whereby they strike each other and produce the expanding sound wave, so that 'the true notion of elasticity is onely, an impediment from compression to a motion that would be freer if the impediment were removed'. The term 'impediment' implies resistance to compression, not restitution from compression. If this is his meaning, then North's conception of elasticity is neither like his brother's nor like Huygens'; rather, we must look to Hobbes's idea of resistance as elastic impact—the endeavour of one moved body contrary to the endeavour of another moved body.[92]

Whatever his source, North rejects his brother's hypothesis that sound results from smaller particles rushing in to fill a space from which gross particles had been driven out. Instead, he echoes Hooke, when he asserts (f. 43) that the main requirement for an audible sound is a vibratory motion that is '*quick enough*' to work upon the elasticity of the air,[93] so that

> when a compression is made, that[,] from the elasticity of the body[,] will compress the next, and so the next, expanding in circle, weakning as it goes, and turning about corners, where is not a force[d] passage; and passeth in to the foramen of the ear thro the ductus of air there, and makes a sence [impression] upon the timpanum, from the quickness of it.

North demonstrates the propagation of this compressional wave by an experiment with a helical spring ('wier worm'), the elastic impact of which excites a longitudinal wave that travels from one end and then reflects back to the other.[94] He also experimentally demonstrates the propagation of transverse

---

[91] See Chapter 2, p. 46 *supra*.

[92] See Chapter 1, p. 11 *supra*. For RN's subsequent treatment of elasticity, see (in chronological order) *NP* 2, pp. 115, 184–185; *NP 5*, pp. 18, 40–41, 87, 150–151; *NP 6*, pp. 11–16, 88, 140; *passim* in 'The Theory of Sounds...' (GB:Lbl) Add MS 32534: ff. 1–82v; and *NP 3*, pp. 97, 156.

[93] See Hooke, 'Lectures de potentia restitutiva', p. 337, where he states that his law explains 'the uniform sound produced by those whose Vibrations are *quick enough* to produce an audible sound, as likewise ... the sounds, and their variations in all manner of sonorous or springing Bodies' (italics mine).

[94] RN utilises the helical spring demonstration in a number of later writings; see (in chronological order) *Cursory Notes of Musicke*, pp. 51–52; *NP 5*, pp. 40, 168; and *NP 6*, pp. 37–38.

waves, direct and reflected, by recourse to 'ropes of great weight' that are 'strained ... where buildings are', although he seems uncertain about the principles of reflection.[95] In spite of his uncertainty, North's kinematic variant of the continuum hypothesis as wave motion in a deformable material medium is a distinct improvement on his brother's variant.[96]

In the first part of his second note (f. 44), North approves his brother's initial definition of tone as equal-timed vibrations that make possible the correlation of pitch and frequency. But in the second part (f. 48v), on his brother's coincidence theory of harmony, he attempts to explain 'the sence of the author more mechanically, and ... more plausible to comon understandings'. To fulfil this purpose, North begins with his brother's explanation of how, after impact, sound is produced by elastic vibrations. To clarify this, he has recourse to a stretched string swinging 'pendulum-wise from side to side'. At the termination of the vibration cycle—or, to use his terms, at 'the instant of the returns' of the string—'sound' is made, because the returning ('crossing') motion of the string compresses the air, which then dilates and sends out an expanding sound wave.

The compressional wave moves 'every thing' within the sphere of it, so that, when excited by the motion of two strings, the vibrations of the air will help 'reciprocally' the vibrations of both. According to North, this is what his brother terms 'a coincidence of pulses'. Hence, if the two strings are tuned in unison, the coincidence happens 'in every returne'. This leads to the problem

---

[95] For a similar experiment made for the same purposes, see Helmholtz, *On the Sensations of Tone*, p. 9. Later on, RN describes the rope experiment in greater detail, revealing that it was made during the rebuilding of St. Paul's Cathedral; see (in chronological order), *Cursory Notes of Musicke*, pp. 51–52; *NP 6*, p. 36 n. 70; and *NP 3*, p. 143. That Cathedral had been destroyed in the great fire of London in 1666, after which Wren was appointed chief architect; see Reddaway, *The Rebuilding of London*. After the fire in the Middle Temple (26 January 1679), RN was involved in plans for rebuilding, so that his acquaintance with Wren probably dates from this time. After DN (2) returned to London in June 1680, he and RN 'usually' went to St. Paul's on 'Saturdays, which were Sir Christopher Wren's days, ... and we commonly got a snatch of discourse with him, who, like a true philosopher, was always obliging and communicative'; see RN, *Life/DN*, p. 238. Hooke also visited the Cathedral nearly a hundred times between August 1672 and December 1680, returning to the site even after that date; see Hooke, *The Diary*.

[96] These improvements also foreshadow RN's later development of his own continuum hypothesis of sound, which, in its 1726 and 1728 versions, states that sound is propagated as pulses moved through the air, resulting when particles of air move in simple harmonic (in-phase) motion, the particles of air acting on each other according to the principles of mechanics; see *NP 3*. For a comparison of RN's and Newton's hypotheses, see Chenette, *Music Theory*, pp. 104–109.

of phase relations, raised by Newton. Like his 1726 answer to Newton, quoted previously,[97] North attempts to solve this problem by recourse to his brother's conception of mechanics—that motions, when opposed, will order themselves so as to balance or equalise the distribution of force. Since the balancing operation 'cannot be the work of above 2 pulses', the lack of conformity at the beginning of the vibrations is 'not perceptible to us'.[98] In short, the ear is insensitive to phase, an assumption held by many later investigators, including Helmholtz.[99]

Finally, there is the problem of dissonance, which Francis North had solved briefly by supposing that audible beats cause dissonance and that the coincidences of some intervals are 'more remote', that is, infrequent. Roger North, however, solves the problem by recourse to elasticity (Hooke's law is implied). Unlike his brother, therefore, he holds that not only 'discording sounds' but 'even the worst of sounds' are periodic ('regular')—that is, have their pulses in proportionable times, although he admits that when the beating is 'to a certein degree', the 'gust doth not relish it'.[100] But he agrees with his brother that musical pleasure derives from functional relations of notes to the keynote ('composed harmony'), not from consonances ordered according to the principle of relative perfection by coincident pulses observed in the fundamental and five secondary tones of the harmonic series ('abstracted harmony').

In his last note (f. 50), he offers a single, but important amendment to his brother's treatment of aerophones. Although North accurately describes that treatment as 'ingenious and new', he disagrees with his brother's explanation of the initiation of vibrations in a flue pipe. He thus enters on the complex subject of the behaviour of edge tone systems, in which frequency depends (in a complicated manner) on the distance from the slit to the sharp edge and on the speed of the air emerging from the slit. Francis North had recognised that the flue pipe is an edge tone system and that trains of vortices ('eddies') are

---

[97] See this Chapter, pp. 102–103 *supra.*

[98] See Appendix A, p. 194 *infra* (RN f. 49).

[99] Helmholtz, *On the Sensations of Tone*, p. 126 *et passim*. According to Taylor, *The Physics of Musical Sounds*, pp. 39–40, 153, for continuous tones in a harmonic relationship, the ear is insensitive to phase; but for intermittent or otherwise varying tones and for tones whose frequencies are close together, relative phase seems to affect the sound heard—as in RN's instance of the 'errors' that spice up the music, quoted in Chapter 2, pp. 75–76 *supra.*

[100] In 'Some Memorandums, concerning Musick' (GB:Lbl) Add MS 32532: ff. 1–26v (see *NP 2*, pp. 36–81), RN states clearly that no note of the scale is discord; rather, discord occurs when a pitch has no orientation to a key or to an accord to a key. For later developments, see (in chronological order) *NP 2*, p. 111; *NP 5*, pp. 17, 83–84; *NP 6*, pp. 86–87; *passim* in 'An Essay of Musicall Ayre...' (GB:Lbl) Add MS 32536: ff. 1–90; *NP 3*, pp. 93–94; and *The Musicall Grammarian*, pp. 167–195.

produced when a jet of air ('film') emerges from a narrow slit. But he did not recognise that when the jet of air impinges on the edge separating the outside from the inside of the pipe, *alternate* vortices will pass along the inside and the outside of the pipe, so that periodic pulses are applied to the air within the pipe [*Figure 3.3*, see also *Figure 2.1 supra*]. That Roger North gleaned this

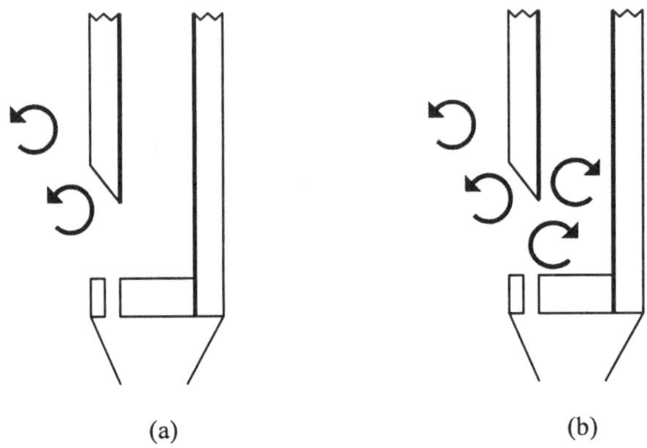

Flue pipe
Initiation of vibration according to (a) Francis North and (b) Roger North

Figure 3.3

principle is clear not only from his description but also from his reference to two experiments, one with a flame that 'seemed to splitt, and flutter both within and without' the pipe, the other with water, where the same thing is seen 'as when a ball is put in and supported by a *ject d'eau*; the weight of it shall carry the stream one way and another', isochronously, 'as a pendulum erected'.

Blowing across an empty or partially empty bottle provides a simple demonstration of a vibration maintained by the edge tone system: the stream of air from the lips striking the edge of the bottle opening supplies the edge tone system, and the bottle acts as a resonator whose frequency is determined by its volume and by the size of the opening. The whistle is an adaptation of the same idea. In the case of a flue pipe, the air flows through a slit between the lower lip of the pipe and a cross barrier—the languid. The stream of air from this slit strikes the upper lip and sets the air column of the pipe into vibration in the following way:

> When the air is first blown through the slit, edge tone vibrations begin to occur. Puffs of air made by the air stream vibrating across the edge enter the air column and produce vibrations of some of its resonance modes, just as plucking a string excites its resonances. One of these, usually the fundamental, will build up most rapidly, and air will begin flowing in and out of the gap between the slit and the edge at the frequency of this mode. The gap between the slit and the edge is called the mouth of the pipe. When air flows into the pipe through the mouth, it pulls the stream from the slit with it. The air stream then blows into the pipe at the same time that the air in the column at the mouth is moving down the pipe. ... Conversely, when the air in the vibrating air column flows out of the mouth, it pushes the air stream from the slit out away from the edge, so that the air stream is also directed out of the mouth. The air stream thus aids the air in the pipe in its motion, and the amplitude of the vibration increases. A standing wave then builds up in the pipe; this wave has a displacement antinode at the mouth since the mouth behaves like an open end.[101]

Edge tone systems, however, were not fully understood until the latter part of the twentieth century, when it was realised that their behaviour resembles that of a mechanical mass-spring combination.[102] Nevertheless, North's conclusion about alternate vortices is significant, because it was a conclusion from a type of experiment that seems to not to have been utilised again until the nineteenth century.[103]

By the time North came to write his note on the flue pipe, he had been acquainted from 1683 or earlier with the organ maker, Bernard Smith, whose organ for the Temple Church was installed in 1688, after a 'battle' between the benchers of the Inner and Middle Temples to decide on the purchase of Smith's organ or an organ built by a competitor.[104] The benchers of the Middle Temple were determined to have 'impartiall' judges chosen to decide the controversy; and in this connection they appointed a committee on 26 June 1685 consisting

---

[101] Backus, *The Acoustical Foundations of Music*, p. 185; see also Roederer, *Introduction to the Physics and Psychophysics of Music*, pp. 119–120; and Taylor, *The Physics of Musical Sounds*, pp. 95–97.

[102] I.e., Hooke's analogue, see this Chapter, pp. 92–93 *supra*; see also Chapter 2, p. 60 *supra*.

[103] See, e.g., Helmholtz, *On the Sensations of Tone*, pp. 91–92, 394–397. RN's use of experiment has yet to be systematically examined. Unfortunately, Korsten, *Roger North*, p. 38, concluded that RN had a 'low' estimation of experiments and, hence, 'consciously or unconsciously followed Descartes'. Had he investigated RN's writings on the philosophy of music, he might have concluded otherwise.

[104] The competitor was Renatus Harris.

of the treasurer and three other benchers for the purpose of requesting that the 'Lord Keeper' determine the matter. But the judges were hardly impartial, since the 'Lord Keeper' was Francis North, himself a bencher and then very ill;[105] and the decision in favour of Smith may have been influenced also by the men whose names appear on the deed of sale and which included Roger North, as well as two gentlemen to whom he was related—Oliver Montagu and Robert Sawyer.[106]

Not long after North began the remodelling of his estate in Norfolk, he contracted Smith to build him an organ.[107] And, when about 1701 the organ was installed in the 'long' gallery—a room over sixty feet in length[108]—North sat at the organ bench with Smith as he voiced the pipes and also tuned them by beats;[109] but North noticed, in particular,

> one nicety [that] was observable in tuning the sharp pipes of mettall in a 15th of an organ. The artist held the pipe in his hand, and very exquisitely pared off some of the mettall, and then set it in place. But it proved not consonant, but very restiff in sound, and varying the beats strang[e]ly. At last in about a minute's time, it fell into the accord as smooth as oyle. I asked the reason of it, and the artist told me the heat of his hand, had warmed the pipe, and till it was cool it did not come to its tone. That heating was such as influenct the spring of the air; for if that in a pipe be more or less rarefyed, the sound will vary accordingly.[110]

Voicing is a process that affects the acoustic spectrum of the tone of the pipe and consists of adjusting the width of the slit, the position of the languid, the position of the lip and so on. It is thus probable that during the decade from

---

[105] FN's illness began before 2 July; in August, he left London for Wroxton Abbey and died there 5 September 1685; see RN, *Notes of Me*, pp. 235–237, 316–317 n.490, and RN, *Life/FN*, pp. 127–130.

[106] For some of the documents, see Macrory, *A Few Notes on the Temple Organ*, pp. 24, 30, 48; and Sotheby's, *Printed and Manuscript Music*, pp. 132–133; see also Wilson, *Roger North*, pp. xviii, 354. For Oliver Montagu and Robert Sawyer, see RN, *Notes of Me*, pp. 188, 229, 296, 313.

[107] See *NP 6*, p. 55; see also Wilson, *Roger North*, p. 56 n.13. In 1786 the organ was moved to the parish church of East Dereham, Norfolk; and in 1875 it was rebuilt by Messrs. Hill and Son.

[108] See RN, 'Cursory Notes of Building', *Of Building*, p. 85:

> It came at last into my mind that a gallery, tho not much used in late fabricks, ... is very delightfull and usefull if it lys comodiously. And if I made the whole of room, from the stair case eastward, which would be about 63 long and 17 foot wide, into one room, and lay the ceiling levell with that of the midle rooms, it would be a good gallery....

[109] For covert references to Smith, see RN, *Cursory Notes of Musicke*, pp. 139–141.

[110] RN, *Cursory Notes of Musicke*, p. 141.

c.1683 to c.1693, North applied some of the insights learned from Smith to improve his brother's explanation of the initiation of vibrations in flue pipes.[111]

II. THE HYPOTHESIS

In his comment, North several times asserts (ff. 42v, 44v, 47v) that, because of the grossness of human bodies, the auditory system ('capacity') has limits for information processing.[112] He also provides a number of instances to make these assertions highly probable. In one instance (ff. 42v–43, 45), he shows that a person will not select information from a 'noisy' background, unless his or her attention is alerted by acoustic signals that are notable—for example, the signals must either 'come up to a certein degree of strength and force', as an explosion, or they must be repeated, as the 'composition' of pendular vibrations that constitute a tone. In another instance (ff. 43v, 47v–48), he shows that the frequencies ('quickness') of sound waves must be 'proportione[d] to our sence', that is, they must fall within the range of human (rather than insect) audibility. Even though the musical scale is limited to tones within that range, North, like Hooke and like his brother, posits a continuum—an infinite series—of frequencies beyond the range of human audibility, for he asserts that 'sense is as minute, and infinite in capacity, as space and time are'.[113]

In North's most significant instance (f. 44v), he shows that human capacity for information processing is limited by

> what we can follow with the motion of any part of our body, either actually or mentally by the help of memory[s], which is the same thing, for wee cannot act in our minds things of which our bodys are uncapable. Therefore when individualls are to pass by in succession, if we can tell [i.e. count] 'em, by nodding or [by moving] the finger, as is usuall in such occasion, so long wee distinguish them. But if they move so fast that wee cannot governe our parts, to note or move to each of them, the distinction of the individualls is lost, and they turne into the appearance of a

[111] For his first complete treatment of flue pipes, see RN, *Cursory Notes of Music*, pp. 88–100; for subsequent treatments, see (in chronological order) *NP 5*, pp. 20–21, 93; ff. 60v, 69, 77–79v in 'The Theory of Sounds...' (GB:Lbl) Add MS 32534: ff. 1–82v; and *NP 3*, pp. 101–102.

[112] For a more recent treatment of this problem, see Miller, 'The Magical Number Seven'.

[113] See Chapter 2, pp. 49–50 *supra*; for Hooke's later theory, see Kassler, *Inner Music*, pp. 137–139, 170–171.

> continued stream, without interruption or distinction, which is a new idea, different from the former.[114]

Since the auditory system at both subconscious and conscious levels computes periodicities, it 'measures'—that is, counts—vibrations arithmetically, in time. Although North explicitly states that counting is the basic (but not the only) operation of the auditory system,[115] this implication had previously been drawn by Hobbes and, then by Hooke, for whom the 'power' of memory, just like the 'power' of hearing to detect pitch, varies with the frequency of vibrations. Hence, memory, an internal sense, is the source of our ideas of time,[116] as Hobbes stated explicitly, when he wrote: 'rough and smooth, like quantity and figure, are not perceived but by the flux of a point, that is to say, we have no sense of them without time; and we can have no sense of time without memory[s].'[117]

If acoustic information is communicated as a composition of periodicities, how is musical knowledge possible? North's answer, hinted in the passage quoted above, is that the origin of musical knowledge, and the condition of its possibility, is the intuition of time that underlies the act of counting. When periodicities are too fast for conscious counting, the auditory system organises the input sequence into larger units—for example, tone (a compound of simple tones).[118] But these units do not merely exist at the moment they are heard, they also endure in memory and relate to what came before and what follows after. Hence (f. 47), the 'duration of time, and the length of space' are equally subject to 'the examination of our senses with the help of memory[s]', so that memory accepts or rejects a unit by a process of

---

[114] Thus (f. 44v), 'reflected specks of light' from falling snow, from multitudes of stars and from a 'fire-stick waived' appear, respectively, as white, as little clouds ('nebulæ') and as 'a stream of fire'; that is, the appearance seems continuous, not intermittent. A number of philosophers before, as well as after RN, have used the fire-stick example. See, e.g., Helmholtz, *On the Sensations of Tone*, p. 170, who used a 'glowing stick' swung round in a circle to fix the limit to the number of beats (intermittent tones) in a second: if the stick executes ten or fifteen revolutions in a second, then 'the eye believes it sees a continuous circle of fire'.

[115] For RN's final version of musical information processing, see Kassler, *Inner Music*, pp. 175–181.

[116] See Kassler, *Inner Music*, pp. 108–112, 124–129.

[117] Hobbes, 'Elements of Philosophy', p. 397; see also pp. 403 and 408.

[118] Hence, the conscious recognition of tone is a spatial, synthetic perception that results from a subconscious temporal analysis. During much of the twentieth century, psychophysics tended to favour an explanation of tone recognition based only on spatial pattern recognition; but Roederer, *Introduction to the Physics and Psychophysics of Music*, pp. 134–135, pointed to an increasing awareness that a neural pulse time-distribution analysis cannot be ruled out.

comparison.[119] For example, if the periodicities are in equal time, 'wee plainely discerne it, and are angry when the order is broke by accelerating and retarding, becaus by what was past wee knew what is coming'.

North's brief treatment of music in terms of disorder that provides a measure of uncertainty forms the rationale for his explanation (f. 45v) of 'the source of all pleasure',[120] for he conceives music as a form of communication. As music moves through time, the attentive listener acquires certain expectations from knowledge by acquaintance—accumulated memories of previous cultural experience. When those expectations are momentarily frustrated, the listener becomes more engaged and consequently more receptive to an emotional or meaningful response. North thus presents a version of what we now call 'information theory',[121] for he stresses the importance of uncertainty in musical communication and the culturally defined nature of musical coherence. Although experience is relative to each listener, beauty (ff. 46v–47) is neither 'a sort of devine stamp', nor the judgment of 'one man's caprice'. North, therefore, rejects the necessary status of universal ideas, as well as the arbitrary, subjectivist bias of clear intelligibility. For him, the subjective grounds of individual judgment must be established on, and confirmed by, the objective grounds of a collective and cumulative experience.[122] Hence, the criterion of excellence is inter-subjectivity—'the

---

[119] See, e.g., Appendix A, p. 184 *infra* (RN f. 46): with the help of memories, 'we can compare, and judg spaces without actuall measuring ... with considerable accuracy'—that is, we can estimate. According to Miller, 'The Magical Number Seven', p. 31, below the number seven humans tend to 'subitize'—apprehend immediately the number contained in a small sample, whereas above that number they estimate. For the limits to 'hearing out' harmonics beyond the seventh, see Roederer, *Introduction to the Physics and Psychophysics of Music*, pp. 137–138 *et passim*.

[120] For his more finished versions of this hypothesis, written shortly after the comment, see RN, 'Cursory Notes of Building', *Of Building*, pp. 10–19, and *Cursory Notes of Musicke*, pp. 142–154. For later developments (in chronological order), see *NP 5*, pp. 33, 131–132; ff. 27, 37v in 'The Theory of Sounds...' (GB:Lbl) Add MS 32534: ff. 1–82v; ff. 8–13 in 'An Essay of Musicall Ayre...' (GB:Lbl) Add MS 32536: ff. 1–90; and *NP 3*, p. 136.

[121] RN's information theory is a transformation of rhetorical theory, whereas, in the 1940s, Claude Shannon, a telephone engineer, based his information theory on a precise definition of the word 'information' as a commodity by which uncertainty is reduced according to statistical rules; see Shannon and Weaver, *The Mathematical Theory of Communication*. For an application of Shannon's theory to music, see, e.g., Pinkerton, 'Information Theory and Melody'.

[122] For the legal basis of RN's reasoning, see this Chapter, p. 108 n.79. For historical changes in the normative bases for truth that seem applicable also to beauty, see Manley, *Convention 1500–1750*, pp. 241–264.

agreement of many' artists of 'best genius and skill' whose works represent a tradition of observation and practice.[123]

[123] For some of these practices, see, e.g., RN, *The Musicall Grammarian*, which includes, pp. 219–278, the first modern history of music—modern, because the author offers a naturalistic (rather than a supernaturalistic) explanation of the origin of music and argues that historical change comes through innovation and technology; see Kassler, 'Roger North', p. 54.

CONCLUSION

With the exception of Roger North, whose numerous manuscripts bear witness to his brother's influence, the reception of the *Essay* remains a subject for study, particularly for the period after the public discovery of its authorship in 1740.[1] But if the *Essay* marks the beginnings of the modern philosophy of music, as defined in this study, then it must share some common ground with the most complete formulation of that philosophy in the work of Helmholtz. In this brief conclusion, therefore, it will be useful to recapitulate, with a broader aim and from a different standpoint, some of the themes that run through the foregoing pages.

To begin with, there is the hypothetical physics that marks the emergence of modern acoustics, a term that was used by various seventeenth-century writers in connection with sound.[2] But it was Isaac Barrow, in 1664, rather than Joseph Sauveur, in 1701, who first suggested that 'a new Part of Mathematics' be 'celebrated by the Name of *Acustics*'.[3] Although this field of investigation predated its name, what distinguished it from scholastic natural philosophy was deliberate experimentation, as in Mersenne's blast-timing experiments or in Boyle's air- pump experiments. In some instances, this new empirical approach led to the quantitative expression of correlations, including those between frequency and pitch of a musical interval, pressure and density of the 'air', stress and strain of a spiral spring, wavelength and length of a pipe.

Nevertheless, there were numerous problems that the new field could not solve until long after the seventeenth century—for example, the problem of auditory localisation and the associated problem of phase relations raised by Newton.[4] For solutions to these and other problems, acoustics had to share jurisdiction with mechanics and elasticity and, ultimately, also with thermodynamics, meteorology, oceanography, the properties of materials, physiology, psychology and so on. But, as Frederick Hunt pointed out in 1978, the characterisation of acoustics 'as a fusional subject nourished by all the sciences' is necessarily a recent concept, 'since some of the subjects ultimately

---

[1] See Chapter 1, p. 21 n.90 *supra*.

[2] For one writer, see Dostrovsky, 'Early Vibration Theory', p. 201 n.6.

[3] Barrow, *The Usefulness of Mathematical Learning*, p. 24. For Sauveur, see, e.g., Dostrovsky, 'Early Vibration Theory', p. 204. For some developments to 1830, see the sources described in Kassler, *The Science of Music*.

[4] See Chapter 3, pp. 98–99 *supra*.

to be involved in fusion acquired their own independent stature only within the last century and a half'.[5]

During the seventeenth century, acoustics shared jurisdiction chiefly with mechanics, which, in the seventeenth century, became the first truly theoretic (deductive) discipline and, with publication of Newton's *Principia*, the model science. But there were different varieties of mechanics, one of which, originating in the two new sciences of Galileo, was conceived as a study of motion against a resistance, whereby motions, when opposed, will order themselves so as to balance or equalise the distribution of force.[6] Introduced by Hobbes to English readers, this variety of mechanics guided the explanations of Francis North, whose central physical hypothesis concerns the action caused by elastic impact—that upon deformation, bodies will oscillate about their positions of equilibrium with a continuous motion to and fro until brought to rest. The law formulated by Hooke in 1678 is fundamental to the mechanics of elastic bodies,[7] which may be experimentally demonstrated, as Hooke did, by mechanical mass-spring combinations.[8] Indeed, it was Hooke who laid the basis for a theory of elasticity as well as for the definition of sound as wave motion in a deformable material medium.

To understand the behaviour of matter in motion, Galileo introduced the pendulum as a physical model. But the pendulum figured so prominently in his work that it has been regarded as a subject of research in its own right.[9] In the course of his research, he discovered the quasi-isochronism of the simple pendulum and the law of variation of its period with length; he also formulated the brachistochrone thesis—that in a circular pendulum the curve of quickest descent is a cycloid. To produce these results, he worked with physical pendula

---

[5] Hunt, *Origins in Acoustics*, pp. 4–5.

[6] By 1800 this mechanics had become 'variational mechanics', in which equilibrium and continuity in change of situation were particularly prominent; see Grattan-Guinness, 'The Varieties of Mechanics by 1800', p. 318.

[7] See Timoshenko, *History of Strength of Materials*, p. 20. See also Helmholtz, *On the Sensations of Tone*, pp. 158–159, who, without naming Hooke, incorporated his law in the statement that

> a simple periodical force, corresponding to pendular vibrations, cannot excite similar pendular vibrations in the elastic body on which it acts, unless the elastic forces called into action by the displacements of the excited body from its position of equilibrium, are proportional to these displacements themselves. This is always the case so long as these displacements are infinitesimal.

[8] This was to become the standard demonstration; see, e.g., Backus, *The Acoustical Foundations of Music*, pp. 11–31.

[9] Ariotti, 'Aspects of the Conception and Development of the Pendulum', p. 351.

(rigid and non-rigid), as well as with ideal pendula.[10] And he also opened up investigations into oscillating systems such as stretched strings, helical springs, columns of air, when he used the pendulum to refute the Aristotelian doctrine, which denied that contrary motions—oscillations—are or can be continuous and regular.

Although physical models are useful, because they offer a way to explore a set of problems too complex for direct analysis, as simulations their predictions are probabilistic. Hence, physical models retain an inherent uncertainty, as Francis North seems to have grasped, since his way of philosophising was probabilistic: it combined hypothetico-deductive with inductive procedures, in which a physical model is the hypothesis or conjecture from which consequences are deduced. In this way of philosophising, verification relies on sampling techniques—repetition of instances (statistical means) and deduction from previous knowledge (analogical means), as in North's use of various kinds of pendula to argue that a stretched string—and, by extension, an air column—is capable of vibrating in more than one mode at once. By 1678 Hooke could substitute a helical spring, since this mechanism is dynamically equivalent to both a pendulum and a stretched string. By the 1690s Roger North could use the helical spring as a model for the propagation of compressional waves in the air and in an organ pipe. And, later on, he could use the same model for understanding human bodies as a 'compage' or engine of springs.[11]

The to and fro motion of these and other oscillating systems takes time; but small time intervals 'eluded science' until Galileo, in his law of falling bodies, treated time not as the course of life—the years, seasons, months, days—but as 'an abstract parameter of a purely physical event', a treatment that enabled him 'to quantify motion in number'.[12] He thus introduced a new concept of time as a mathematical continuum and, hence, as an independent and measurable variable.[13] With this new concept, the observer is placed *outside* the temporal framework, since time can be represented by a unidirectional straight line and coordinated with spatial facts similarly

---

[10] In an ideal pendulum, the thread of the pendulum becomes the radius of the circle and the bob, a moving point. For Hobbes's interpretation, see Kassler, *Inner Music*, pp. 54–58.

[11] See Kassler, *Inner Music*, pp. 168–169.

[12] Gillispie, *The Edge of Objectivity*, p. 42; see also p. 90.

[13] Cassirer, *The Philosophy of Symbolic Forms*, vol. 2, p. 138, gave precedence to Johannes Kepler, in whose three laws of planetary motion, time appears as the fundamental variable—'that uniformly changing magnitude to which all un-uniform change and motion are referred and by which they are measured'. According to Cassirer, this is the 'ideal, purely logical significance' of the time concept, which, 'imbued with the concept of function', from now on 'appears as one of the most important applications and expressions of functional thinking'.

represented.[14] In the second part of his *Essay* on the mechanics of music, Francis North utilised the new concept, when he treated pitch as a continuum of frequencies, those within the range of human audibility being measurable and represented either numerically, as 1, 2, 3, 4, etc., or spatially, as vertical dots on a horizontal time grid ('Scheme' *Figures 1.B* and *1.A*). But in the third part on the semantics of music, he introduced a different conception of time as the measure of musical motion, thereby placing the observer *within* the temporal framework. For the attentive listener, therefore, time cannot be unidirectional—that is, passing from the past into the future—because, in experienced time, there is inclusion of the past in the present, as well as anticipation of the future from the present. Hence, North represented the attentive listener as a measuring instrument that traces movement of music through time like a swinging pendulum.

This dual conception of time as objective process and as subjective experience provides a hint of the later separation of the philosophy of music into two distinct sciences, acoustical and musical.[15] As stated at the outset of this study, it was Helmholtz who sought to reconnect the two sciences by reducing the phenomena of hearing to resonance—the synchronisation of one periodicity or rhythm with another.[16] To effect this reduction, he modelled the membrane of the eardrum as a system of horizontally stretched strings, the basilar membrane as a damped system of such strings, and the membranous connection as the fulcrum to the pressure of the fluid against these strings.[17] Unknown to Helmholtz, however, a similar type of reduction may be traced to the 1675 lecture of Hooke, where the eardrum is conceived as a stretched string that vibrates in phase with incoming vibrations, and to the 1677 *Essay* of North, where the auditory nerve is conceived as a stretched string that vibrates simultaneously in more than one mode.[18]

---

[14] For the old, Aristotelian doctrine of time as the continuous transformation of potentiality into actuality, as well as for the new, Galilean concept, see Burtt, *The Metaphysical Foundations of Modern Science*, pp. 94–96 *et passim*.

[15] See Kassler, *Music, Science, Philosophy*, pp. 165–200.

[16] In synchronisation, *N* cycles of one periodicity match *M* cycles of the other, *M* and *N* being integers (the commonest mode is 1:1).

[17] Helmholtz, *On the Sensations of Tone*, pp. 135, 146, 148.

[18] In 1683 J. G. Duverney published the most detailed theory of auditory resonance, see Duverney, *Traité de ... l'oreille*, which takes its starting point from the work on elastic impact and the mechanics of vibration of Mariotte, *Traité de la percussion ou choq des corps*. Duverney conceived the eardrum as an elastic body that conforms to the vibrations of sonorous sources: when stretched, it vibrates faster and when relaxed, it vibrates slower. Other parts of the internal ear react to these motions by resonance, including the *lamina spiralis* of the cochlea, conceived as a helical spring, the larger parts of which form the slowest vibrations and the narrower parts, the quicker vibrations. His book was noticed in the Royal Society's

In these oscillating systems the basic form of motion is pendular—simple harmonic motion—for it is peculiar to the simple string pendulum and cannot be further analysed. Resonance thus provided Helmholtz with a reason 'why an originally simple periodic vibration of the air produces a sum of different sensations, and hence also appears as compound to our perceptions'.[19] To analyse the compound into its sensory constituents requires a special act of attention, because, for Helmholtz, the 'common property' of all sense perceptions is 'the apprehension of compound aggregates of sensations as sensible symbols of simple external objects, without analysing them'.[20] Helmholtz, therefore, distinguished synthetic from analytic perception.[21] In synthetic perception, simple tones corresponding to pendular vibrations of a compound motion of the air 'fuse into the whole mass of musical sound', whereas in analytic perception those simple tones can be recognised 'without any other help than a proper direction of attention'.

For Helmholtz, all attributes of a musical tone are compounded out of the sensations of several simple tones ('partials'), and the compound is formed in experience, because the sense perception of several simple tones habitually occur together. This is his principle of conscious decay under habituation, mentioned at the outset of this study.[22] Although the bare sensory pattern, as directly dependent upon the auditory stimulus, is a 'perception', a pure perception is comparatively rare, because Helmholtz supposed it nearly always is supplemented and modified by an imagined increment, dependent upon memory and induced by what he called 'unconscious inference'. This imagined increment makes a perception into an intuition (*Anschauung*), or synthetic perception, which literally is the perception, since by intuition objects are correctly perceived and truly identified. If the sense impressions are entirely lacking and the listener has only the imagined equivalent of the intuition, then the experience may be called a presentation (*Vorstellung*). Hence, the key to

---

Transactions (1683) [No. 149, 10 July 1683], 13: 259–264; see also Birch, *The History of the Royal Society*, vol. 4, p. 205.

[19] Helmholtz, *On the Sensations of Tone*, p. 148.

[20] Helmholtz, *On the Sensations of Tone*, p. 368.

[21] Helmholtz, *On the Sensations of Tone*, pp. 62–63, where Alexander Ellis commented that

> Prof. Helmholtz uses Leibnitz's terms *percipirt* and *appercipirt*, alternating the latter with *wahrgenommen*, and then restricting the meaning of this very common German word. It appeared to me that it would be clearer to an English reader not to invent new words or restrict the sense of old words, but to use *perceived* in both cases, and distinguish them (for *percipirt* and *appercipirt* respectively) by the adjuncts *synthetically* and *analytically*....

[22] See Chapter 1, pp. 1 and 3 *supra*.

sensory perception lies in the intuition—synthetic perception—that involves both auditory stimulation and unconscious inferences.[23]

The first hints of this doctrine of unconscious inference appear in the philosophy of Hobbes and in some writings of Hooke.[24] But the doctrine may be gleaned not only in Roger North's comment on his brother's *Essay* but also from his descriptions of Hooke's toothed-wheel experiment, cited previously, and from which he concluded that, when the vibrations are faster than we can count, they appear in 'imagination' as a compound.[25] To assist the attention in analysing the compound, Francis North devised a graphic method to demonstrate its multiple frequencies (see *Figure 1.A*), whereas Helmholtz could use a different method—Fourier analysis[26]—to show that: '*Any given regular periodic form of vibration can always be produced by the addition of simple vibrations, having pitch numbers which are once, twice, thrice, four times, &c., as great as the pitch numbers of the given motion.*'[27]

Initially, mechanisms like the pendulum contributed to the conclusion that musical instruments have a constant natural period independent of the effects of amplitude.[28] But this principle of isochronism—equality of timing—led Francis North to the further conclusion that information processing in the auditory system is determined by the isochronous nature of coupled systems, so that pitch is ordered by the equality of small time intervals and metre, by the equality of rhythm. In this way he established equality as the norm for musical information processing at both levels of the auditory system—subconscious for pitch and conscious for metre. Indeed, as he hinted, equality is 'natural' to the auditory system, since, for example, the temporal perception of rhythm is bound up with the process of attention, which North modelled on the oscillating pendulum. But Roger North went further, for he supposed (f. 47)

---

[23] For the doctrine of unconscious inference, see Helmholtz, *On the Sensations of Tone*, pp. 62–65, 289–290, 368–369; see also Boring, *Sensation and Perception*, pp. 399–436.

[24] See Chapter 3, p. 118 *supra*; see also Kassler, *Inner Music*, pp. 110, 129, 133–139 *et passim*.

[25] See Chapter 2, p. 72 *supra*. Chenette, *Music Theory*, in his otherwise admirable study, misinterpreted RN's synthetic perception as an innate idea. He was more accurate in stating, p. 194, that for RN, 'a musical sonority is a human experience with dynamics in a physical world of vibrating forces given a temporal dimension by memory'.

[26] I.e., the analysis named after J. B. J. Fourier, who in 1822 demonstrated that a finite and continuous periodic motion can always be decomposed into a series of simple harmonic motions of suitable amplitudes and phases. Subsequently, G. S. Ohm applied this analysis to music; see Beyer, *Sounds of Our Times*, pp. 44–45.

[27] Helmholtz, *On the Sensations of Tone*, p. 34; for Fourier analysis, see also pp. 35 and 229.

[28] According to Cannon and Dostrovsky, *The Evolution of Dynamics*, p. 6, up to the middle of the eighteenth century, 'isochronism and simultaneous crossing of the axis always imply the pendulum condition', so that motions not having these properties were called 'irregular' and were not treated.

that 'in all wee doe wee affect equality and regularity'; and he later assumed that, for living creatures, isochronous motion is a natural necessity, because 'the laws of all pendulous or springy vibrations require equality of time, as moderne philosofy hath with clear reason, and with constant experiment, proved'.[29]

What is it, then, that transforms regular vibrations—equalities—into a pleasing sound?[30] In 'Some Notes' on his brother's *Essay*, Roger North answered this question by recourse to his test of comprehensibility, when he distinguished between tone, a periodic vibration, and noise, a non-periodic vibration, to show that regularities are better understood than irregularities.[31] As for the more complex question—what constitutes pleasing music?—his answer (f. 49v) was that pleasing, when applied to 'musick in consort', denoted 'some vniforme and regular mixtures of pulses in diverse proportionate measures of time'. It was but a short step from this conclusion to posit equality as the norm for music itself, a step that Roger North took sometime between c.1710 and c.1715, when he explicitly stated that 'all musical time' is measured by 'equality'—duple metre (common time)—and that triple metre is 'but a manner of breaking it without infringing the stability of the measure and the equality of it'.[32] In short, metre is structurally isomorphic with pitch.

It would seem, then, that equality of timing is the fundamental premise of the modern philosophy of music that emerged from 1677. Hence, we might ask: how can modern philosophers of music guarantee that, for example, two time intervals are equal? The answer is: they postulate that systems, if uninfluenced by accelerating or retarding forces, will move in such a way that equal distances spell equal times. It is this postulate that enables them to construct time after certain abstract mathematical patterns, as well as to apply their constructions to oscillating systems and to music itself. Although the postulate of equality thus functions as a constructive principle, it also has a second function as a regulative principle, since it *restricts* the way in which systems can be conceived and described. Perhaps, Newton grasped this second

---

29 See f. 16 in (GB:Lbl) Add MS 32531: ff. 1–41v, which includes RN's animadversions on two tracts by a musician who had been in the service of James II. For a description of this manuscript, with an analytical index of its contents, see *NP 6*, pp. 43–152.

30 It seems that before 1677 none of the early modern philosophers addressed this question; see Cohen, *Quantifying Music*, pp. 211–212 *et passim*.

31 See Appendix A, pp. 184–192 *infra*, (RN ff. 45, 46, 47, 48). The implication, developed in his later writings, is that in information processing—discernment by the ear and by the intellect—equality of timing is required for knowledge, else there is confusion and, hence, ignorance.

32 See f. 17 in (GB:Lbl) Add MS 32531: ff. 1–41v. For the application of equal-time theory to prosodic construction, see Omond, *A Study of Metre* and *English Metricists*; Fussell, *Theory of Prosody*, p. 82 *et passim*; and Kassler, 'Joshua Steele'.

function, when he sought to undermine the in-phase supposition of the coincidence theory of harmony.[33]

Newton's criticism notwithstanding, the in-phase supposition was a reasonable one, since it derived from investigations into the nature of mechanical oscillating systems, including the auditory system so understood. From our present standpoint, there are several points to note about such systems. First, they have two essential requirements: restoring forces (usually elastic) that tend to return the system to its former state undisturbed; and accelerating forces (inertial mass) that tend to delay the return and subsequently to carry the system beyond the equilibrium position to maintain oscillations. Second, a number of special properties characterise their oscillations. One is that the amplitude of oscillation is completely independent of the frequency as well as a discrete set of frequencies for which the system will oscillate. Another is the property mentioned so frequently in this study—that oscillators can and generally do oscillate with constancy of period. These two properties define an oscillator's *linear* behaviour, according to which the outcome produced by a given increase in input does not vary and, hence, is predictable. But there is yet another property—hinted at by both Francis and Roger North but otherwise scarcely noticed during the seventeenth century—whereby a small change in the input of energy to a component of an oscillator triggers a major change in what the component does (how it oscillates). This property, which makes prediction a problem, is known as the dependence of an oscillator on initial conditions (how it comes to oscillate), and it defines such *non-linear* behaviour as the onset process for mechanical oscillators where strict periodicity does not obtain.[34]

Helmholtz himself recognised a number of instances, both linear and non-linear, where strict periodicity does not obtain.[35]. For example, he pointed out that tempered tuning of musical instruments prevents 'an accurate fulfilment of the conditions under which alone the resulting motion of the air

---

[33] See Chapter 3, pp. 98–100 *supra*.

[34] A system is linear when the displacement is proportional to the restoring force, so that the oscillating motion is simple harmonic (pendular) motion, whereas it is non-linear when the displacement is not directly proportional to the restoring force. See, e.g., Wiener, *Cybernetics*, p. 109, for a brief treatment of the difference between the linear and non-linear theory of organ pipes. According to Winckel, *Music, Sound and Sensation*, p. 25: 'Strictly speaking, there are no exactly periodic events in our life on earth, since every technically generated vibration must have been switched on at a finite point in time, leading to distortion'.

[35] Helmholtz, *On the Sensations of Tone*, pp. 152–159, argued that the middle ear (eardrum and ossicles) cannot be presumed linear when a certain intensity of stimulus overloads it, thereby causing the middle ear to generate combination tones. With this explanation, he separated the (non-linear) theory of combination tones from the (linear) theory of beats; see Maley, *The Theory of Beats*, p. 133.

can be *exactly periodic*'.[36] Nevertheless, he acknowledged that temperament facilitated greater harmonic freedom by allowing composers to move through an increasingly large number of different harmonic relationships, as well as to modulate from one key to another. But he expressed concern that, 'in many modern compositions', frequent recourse to 'dissonant chords', as well as to 'restless modulation' threatened 'to destroy the feeling for tonality' by obliterating 'artistic connection'—the internal musical order among all the notes of a composition that makes their relation to the keynote as perceptible as possible. According to Helmholtz, therefore, these were 'unpleasant symptoms' for the future development of the art, since, if unchecked, they would lead to a loss of musical coherence, as it was defined between 1677 and 1877.[37]

As is well known, new conceptions of musical coherence began to emerge after 1877, first gradually and then, from about 1908, with increasing momentum. Instead of relying on a tonality ('keynote') of a major or a minor diatonic scale of seven notes plus five 'accidentals' and a tempo ('time') measured by symmetric periods, the new conceptions ordered pitch and metre in a number of different ways—for example, by recourse to unusual scales, including a chromatic scale of twelve equally tempered notes, and by overlapping speeds to produce constant changes of metre or dislocations in rhythm.[38] Nevertheless, these new conceptions were evolutionary, not revolutionary for two main reasons. First, they were responses to, and even commentaries on, the previous conception, as, for example, by making tonality fluctuating and, hence, ambiguous; by combining several tonalities (polytonality); by avoiding tonality altogether; even by emphasising one of the twelve notes as a 'tonal centre' so as to create a feeling of tonality. Second, the new conceptions were tied to traditional musical instruments and, hence, to the harmonic series—though perhaps with less reliance on the fundamental and more reliance on the higher secondary tones—until, in the 1950s, advances in electronic technology allowed the generation of a range of different sonorities, including noise, as elements for creating coherence. Nevertheless, the

---

[36] Helmholtz, *On the Sensations of Tone*, p. 30 (italics mine); see also pp. 319–330, for the supposed advantages of just intonation over equal temperament. From the standpoint of Winckel, *Music, Sound and Sensation*, pp. 127, 161, too much has been made of the theoretical differences between tuning systems—for hearing, these differences are slight, because 'the pitch differentiation sense in the ear is not so well developed that it can determine a pitch exactly'—the very point made by Brouncker and FN; see Chapter 2, p. 80 *supra*.

[37] See Helmholtz, *On the Sensations of Tone*, pp. 319, 327, 349.

[38] See, e.g., Biggs and Orzel (eds.), *Musically Incorrect*, pp. 31–32, where the composer, Milton Babbitt, pointed out that, for serial music, the structural isomorphism between a pitch collection and a temporal collection of 'time-points' is 'simply' higher or lower than, earlier or later than.

conception of musical coherence that emerged from the seventeenth century has continued to co-exist with the more recent conceptions, so that in 1942 one composer went so far as to assert: 'Tonality is a natural force, like gravity.'[39]

From the standpoint of the philosopher, tonality is not a natural force but a concept—a highly complex one of course, for it defines a particular kind of musical reality and is universal for that reality. But the musical reality called 'tonality' is not a static but a dynamic quality, because it changes as new discoveries are made. Hence, concepts do not emerge fully formed; instead, concept formation is an historical process that begins as a new programme for research, as in Francis North's *Essay*, and reaches completion only much later, as in Helmholtz's *Tonempfindungen*. In the words of Ernst Cassirer,

> ...it remains true, even in highly developed knowledge, that each newly acquired concept is an attempt, a beginning, a problem; its value lies not in its copying of definite objects, but in its opening up of new logical perspectives, so permitting a new penetration and survey of an entire problem complex. Thus while among the basic logical functions the judgment closes and concludes, the concept, by contrast, has essentially the function of opening up. ... In this sense a concept can be effective and fruitful for knowledge, long before it is itself exactly "defined," i.e. carried to a complete and definitive determination.[40]

Since this study is limited to the beginnings—the opening up of new logical perspectives, perhaps it, too, will be fruitful for future research into the historical formation of the modern philosophy of music.

---

[39] Paul Hindemith, quoted by Searle, *Twentieth Century Counterpoint*, p. 60.

[40] Cassirer, *The Philosophy of Symbolic Forms*, vol. 3, p. 306; see also Kassler, *Music, Science, Philosophy*, pp. 61–82.

APPENDIX A: EDITIONS

EDITORIAL NOTE

In this appendix, editions are provided of four texts discussed in the foregoing study: the *Essay* of Francis North, the synopsis of the *Essay* by an anonymous individual, and the critical comments on the *Essay* by Isaac Newton and by Roger North. Of these four texts, the originals each differ in format. The first and second are printed—one as a book, the other as a short notice in a periodical publication. The third and fourth are in manuscript—one, a private letter, the other, an incomplete set of notes. All the texts conform to seventeenth-century practice which differs in some respects from modern practice. Hence, the following principles have been applied in editing the texts.

Pagination/foliation. In the editions, the original pagination or foliation of each text is indicated in bold type at the appropriate places within a text. This original pagination/foliation is used for all references not only in the foregoing study but also in the footnotes to the edited texts. To determine the pagination in Francis North's *Essay*, ten copies were consulted for the present edition.[1] In all these copies, the final page of text is incorrectly numbered '25'.[2] In most copies the engraved plate containing Tab. i and ii (*Figure 1.A*), which was issued with a fold at the bottom, is cropped and bound uniform with the size of the other leaves, being placed with the 'Scheme' after the last page of text.[3] In placing the engraved plate and the 'Scheme' differently in the edition, I treated the unpaginated front matter as follows: page i = title page; page ii = blank; pages iii and iv = epistolary introduction; page 1 = blank; page 2 = engraved plate (*Figure 1.A*); page 3 = 'Scheme'; page 4 = blank. (From page 5 the *Essay* is paginated.)

---

[1] (GB:Lbl) 557*.c.20(1), 557*.c.21(1), 557*.c.25(1–2), 432.c.15(8); (GB:Cu) MR 470.d.65.1 and Adams 7.65.18[6]; (GB:Cjc) AA6.22; (US:Dlc) ML3805.A1.G9 and ML3805.A2.G9; (A:Nno) Kassler Library. (GB:Lbl) 432.c.15(8) was owned by Joseph Banks, who bound his copy with the Cutlerian lectures of Hooke (two of which are cited in this study).

[2] (GB:Cjc) AA6.22 has a number of corrections in a contemporary hand, including a correction on the last page, where the '2' has been altered to '3'. From other manuscript annotations in this copy, it was probably owned by a branch of the North family.

[3] The exceptions are (US:Dlc) ML3805.A2.G9, which lacks the engraved plate and the 'Scheme'; (GB:Lbl) 557*.c.21(1), where the engraved plate and the 'Scheme' are misbound with another tract; (GB:Cjc) AA6.22 and (GB:Lbl) 557*.e.25(1-2), which have the engraved plate and the 'Scheme' placed before page 5.

Spelling. Spelling is retained, including use of 'then' for 'than'. Where necessary for sound or sense, a letter is added in square brackets (e.g., 'breath[e]'). The use of apostrophe 's' for possession is standardised; but the use of apostrophe 't' for past tense or past participles is retained (e.g., 'expres't'). When the apostrophe 't' is not used (e.g., 'unmixt'), it is added only when sense requires it (e.g., 'plact' becomes 'plac't').

Contractions. All abbreviations are expanded; e.g., ♯ becomes sharp, ♭ becomes flat, *sc.* and *scil.* become namely. Abbreviations still in common use are retained, but '&c.' is given its modern form, 'etc'. While it is possible that Newton's 'pag.' and 'lin.' may be abbreviations for the Latin *pagina* and *lineamentum*, they are expanded in the edition as 'page' and 'line'. Fractions are retained for proportions only, so that '1/2 note' (*Essay* p. 25) becomes 'half note'.

Punctuation. All full stops are dropped from section headings in a text. For lists, punctuation is added without comment. Where punctuation is altered to conform more to modern practice, the guide has been clarity of meaning, especially when breaking a long sentence at a colon or semi-colon or replacing a colon with a full-stop, a semi-colon or a comma. All other editorial punctuation is given in square brackets.

Capitalisation and Italicisation. Capital letters mid-sentence, apart from proper nouns, are changed to lower-case. Words in italics are changed to roman except for a foreign expression (e.g., '*in infinitum*'), a book title, and underlined passages.

Errors. When homophones are confused (e.g., to/too, of/off), they are emended silently. In the case of Francis North's *Essay*, where the original manuscript is no longer extant, some of the errors may be those of the printer, who issued the *Essay* with a slip cancel headed '*Mistakes.*' It reads as follows:

> Page 9. l. 18. for *line* read *time*: l. 25. for *fifth and fourth* read *five and four.*
> Pag. 24. l. *penult.* for 5/2^{d} *and* 3/7th ♭ read 2^{d} *and* 7th ♭[.]
> Pag. 30. l. 13. for *Octave to revers* read *Octave or reverse.*
> Pag. 15. l. 12. *so regular* dele *so*.
> Pag. 26. It must be observed that the *Monochords* are not divided as they ought to be; the *Reader* therefore is to regard only the figures, and not the length of the Lines.

The sentence following 'Pag. 26' in the slip cancel is inserted in the appropriate place in the text (see p. 158); and the other errors are corrected without comment. Errors not indicated in the slip cancel, as well as errors in other texts are identified in the notes.

Illustrations. In the *Essay* Francis North uses the treble and tenor clefs for the figure in his 'Scheme' (*Figure 1.B*). In the edition these are transcribed in the treble and bass clefs. The two other figures (*Figures 1.A* and *1.C*) are reproduced from a copy of the *Essay* in the Kassler Library (A:Nno). Captions are added to all three figures The third figure (*Figure 1.C*) constitutes page 26 of the *Essay*, the format of which is difficult to read. But the page also has a number of apparent errors. Hence, at the end of the edition, a clearer representation, with corrections, is provided in additional pages (see *Figure 1.D*, pp. 167–170). The corrections in these pages are as follows: Second page (p. 168) line one, read 360/16 for 360/15; and line six (false fifth), read 270/19 2/7 for 270/19 7/2. Fourth page (p. 170), read 'D' for 'Da'; and read ♯ for ♭ for both the sharp and the flat key. Since (p. 170) North starts 'Ut' on C, F and G, the solmisation syllables have been re-arranged to clarify this. In the edition of Newton's letter, a caption is added to the figure (*Figure 3.A*), which is reproduced from the redrawing included with the letter's first published transcription (see Miscellaneous below). In Roger North's manuscript, the figures were drawn in the margin as small, rough sketches. In the edition all figures are redrawn with captions added. But two of North's figures were so sketchy that they are redrawn not as copies but as clarifications of his meaning (see *Figures 4.C* and *4.D*).

Footnotes. In addition to indicating errors in a text, these notes serve a number of different purposes, including cross-references within a text or from one text to another, interpretation of difficult passages, identification of a person or a source. Since Roger North's text includes revisions in the left-hand margin, these are placed in the notes, since it is not always clear where they might be inserted in his text. The majority of the notes provide the meaning of terms that are not in current usage or whose usage is different from that current. Some of the notes give the meaning from an equivalent word (e.g., '*bigness*, i.e., length'), whereas some give the meaning and dates from the *Oxford English Dictionary* (OED). On one occasion Roger North uses a word for which the OED can give no other source. This is termed a 'Northism', even though North's usage may simply reflect contemporary colloquial practice. Other notes relate to scientific, technical and musical terms. All these terminological notes are intended to assist readers from different disciplines, as well as those who are not accustomed to reading seventeenth-century texts. Readers should note that two terms—'sense' and 'memory'—occur in some of these texts with multiple meanings. In the case of 'sense', this may denote a sense faculty, a sense perception or a meaning; in the case of 'memory', this may denote a mental faculty, a past experience (remembrance) or past experiences (memories). No footnote comment is added for these terms, although an editorial '[s]' is inserted when the term 'memory' has the meaning of memories.

Miscellaneous. Newton's letter to John North has been transcribed and published by (1) John Addington Symonds, Jr., 'A Letter of Sir Isaac Newton', *The Academy* (December 11, 1869), pp. 73–74; (2) Henry H. Dale, 'A hitherto unpublished Letter of Isaac Newton', *Hinrichsen's Musical Year Book 1945–46*, vols. 2–3, pp. 400–403; and (3) H. W. Turnbull, 'Letter 206', *The Correspondence of Isaac Newton* vol. 2 (Cambridge, 1960), pp. 205–208. The second transcription also includes a facsimile of folio 2 of the original letter. In its first appearance in *The Academy*, Symonds submitted the transcription of the letter with a covering note, dated Clifton, Bristol, 3 December 1869, in which he stated that the letter was then the property of 'my sister-in-law, Miss North, of Hastings Lodge, Hastings'. This would be Marianne North, flower painter, traveller and sister of Symonds' wife, Janet North Symonds. Since the original letter has brief annotations not only in the hand of Roger North but also in the hand of his eldest son, Roger, and his great-great grandson, Frederick, it seems that the letter had passed to Marianne North through the male descendants of Roger North—his son, Roger North; his grandson, Fountain North; his great grandson, Frederick Francis North; and his great-great grandson, Frederick North, MP for Hastings and father of Marianne and Janet.[4] Symonds also mentioned that in order 'to give further security for its genuineness, I propose to place the original in the hands of General Sir Edward Sabine, Bart., for the inspection of the members of the Royal Society'. The letter was then returned to its owner until the outbreak of World War II, when the second person to transcribe it, Henry H. Dale, noted that the Royal Society 'recently received, on indefinite loan from its present owner, a hitherto unrecorded autograph letter of Isaac Newton'. At that time the owner was Roger North, father of the present owner, Thomas Frederick North, who repossessed the letter.

---

[4] See Marianne North, *Recollections of a Happy Life*, pp. 1–5, also for information about Fountain North's two sisters and the family of Frederick Francis North.

1

FRANCIS NORTH

A PHILOSOPHICAL ESSAY OF MUSICK DIRECTED TO A FRIEND

[(p. iii)]
Sir,
Having often tyred you with discourses of several parts of musick, wherein you desired to receive satisfaction, without obtaining the effect I proposed to my self; I reflected upon the saying, that no man understands that which he cannot so express as to make another understand it,[1] and doubted whether the hypothesis I entertained were not founded upon errours, which I had swallowed without any due examination. I resolved for tryal of it, to trace my notions as near their principles as I could, and set them down in method, that I might discover whether they were well deduced from one another, and from such experiments as I had in memory. Having done this I send them to you, as the severest judge I know. It has not been your custome to flatter me, and now I desire your censure, if I may impose it upon you to give it in writing, wherein you will be obliged to quote what I say truly, before you object to it, and to stand to those allegations and inferences you will set down for your own, in both which respects I had great disadvantage in discourse. I shall prevent you in condemning the stile, for being too contracted[2] and obscure; and you must take the fault to your self when you know it does not proceed from the aversion I have to philologie,[3] but that I thought it would be labour lost to enlarge, when writing to so great a philosopher and musician as you are, I cannot fail of being understood. **[(p. iv)]** In other things I expose it to you; and if I shall receive plain confutations, it will not much disappoint or mortifie me. For I have hardly in my whole life made one set experiment upon this subject, nor know many of the instruments I have spoken of. But what I heard from others, or

---

[1] FN may have adapted this saying from what he had learned as a pupil at the King Edward VI free grammar school, Bury St. Edmunds, since, according to John Aubrey, FN's 'tutor (as his Lordship told me) made him properly to understand what he read to him before he left him, so no doubt was left upon him'. Aubrey supposed that this practice was the basis not only for learning grammar but also for learning 'all the arts'. See Stephens (ed.), *Aubrey on Education*, pp. 56, 163.

[2] *contracted*: OED *ppl. a.* 5.b. *fig.* condensed, concise (1595–1818).

[3] *philologie* = *philology*: OED *sb.* 1. love of learning and literature (general sense now rare c.1386-1892).

occasionally observed, I laid up in mind, and established this systeme, which I shall be glad to lose, if I may improve my knowledge by your discoveries. This may seem strange to you, considering how fond men are of their own inventions. At the worst, I hope for this fruit of my pains, that you will make such experiments as your curiosity will suggest whilst you read this, and let me know the success of them. Though I may thereby be driven out of this hypothesis, perhaps I shall receive light towards another, which you may believe I shall take as much pleasure in making, as you, or any body else can to destroy.

I am, Sir,

Your most faithfull friend
and servant.

[(p. 1 is blank)]

[(p. 2)]

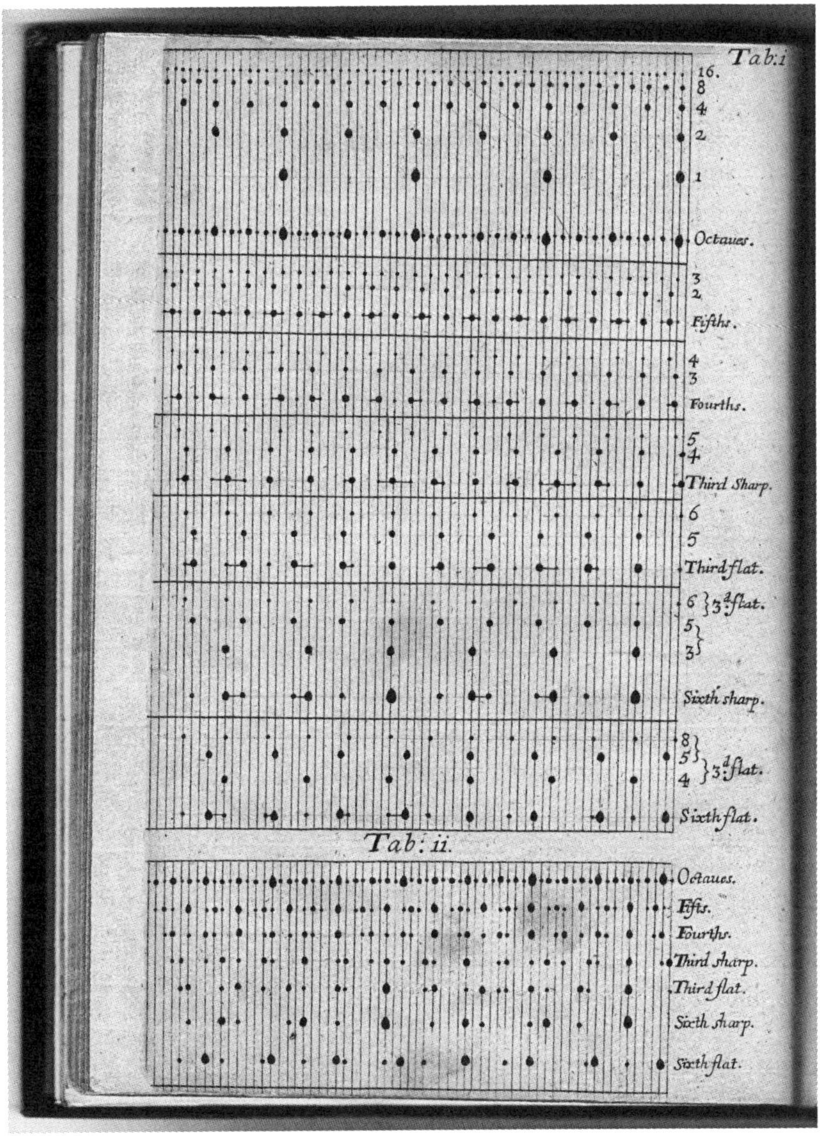

Graphs showing the periodicity patterns of consonances in relation to the fundamental tone as a function of time

Figure 1.A

[(p. 3)]

A
SCHEME
WHEREBY THE SYMMETRY OF THE PULSES RANGED IN MUSICAL PROPORTION APPEARS

In the first Table [(p. 2) *Figure 1.A*, Tab. i],[4] the pulses comprehended in the several chords,[5] are first set down in several lines, and then in one line with couplings, for the more easie observing the regular distances[6] of the by-pulses[7] on each side of the coincidence.

The sixth sharp and flat[8] have their reverses,[9] being the third flat and sharp.

In the second Table [(p. 2) *Figure 1.A*, Tab. ii], the several chords are in several lines without couplings.

It is worth observing, that the close in musick, being the most pleasing, and full sound, which comprehends the fifth and third sharp, and eighths, hath the pulses in the proportion of these numbers, 1, 2, 3, 4, 5, 6; as appears by this figure [*Figure 1.B*]

First six notes of the harmonic series arranged as a major chord over a fundamental bass

Figure 1.B

---

[4] In the last segment of Tab. i [*Figure 1.A*], for '3d. flat' read '3d. sharp'.

[5] *chords, chord*: OED *sb.*[1] 1. *obs.* = accord *sb.* 4. agreement of musical sounds (c.1475–1636); see also FN pp. 9, 10, 15, 17, 18, 19, 20, 21, 22, 23, 24, 25, 28, 29, 30, 31, 32, 35. For specific meanings as *(a)* individual tones in a complex tone, *(b)* overblown tones and *(c)* flageolet tones, see Chapter 2, pp. 50–51, 62–68 *supra.*

[6] *distances, distance*: OED *sb.* III. 11. *transf.* the extent or 'space' of time between two events; an interval, intervening period (c.1384–1871); see also FN pp. 8, 9, 10, 33.

[7] *by-pulses; by*: OED *a.* 2. *fig.* b. of secondary importance (c.1050–1857); see also FN p. 10 and intermediate pulses, FN pp. 8, 9, 10.

[8] *sharp and flat*, i.e., major and minor *(a)* applied to intervals as here, see also FN pp. 9, 11, 23, 24, 25, 30 or *(b)* applied to modes ('keys'), see FN pp. 19, 23 and *Figure 1.C.*

[9] *reverses*, i.e., inversions; see also FN pp. 11, 27 and reverse, FN p. 30.

[(p. 4 is blank)]
[p. 5]

OF SOUND

My design being to explain the nature of musick it will not be improper to enquire first, concerning the cause of sound. And here though I cannot hope to give full satisfaction, because our senses want subtilty to discover the motions of the minute parts of the air whereupon it [sound] depends, yet perhaps I may assign a possible cause for the producing of sound, the knowledge whereof will give some ease to the wondring philosopher. And if I have the good fortune to shew a probable cause, the curious will acquiesce from[10] any farther enquiry, which is all the benefit they can have by speculations of this nature.

The phænomena of sound which I think considerable are these: 1. It may be produced in the Torricellian vacuity. 2. It causes motion in solid bodies. 3. It is diminished by interposition of solid bodies, and 4. If the bodies interposed are very thick, its passage is wholly obstructed. 5. It seems to come to the ear in strait lines when the object is so scituated that it cannot come in a straight line to the ear. 6. When the air is not in motion its extent is sphærical. 7. When there is a wind, the sphære is enlarged on that part to which the wind blows, and diminished on the contrary part. 8. That it arrives not to the ear in an instant but considerably slower than [light to the organ of] sight. 9. That it comes as quick against the wind as with it, though not so loud nor so far.

There are many other observable appearances, and more will daily occurre to the curious upon experiments, but the consideration of these qualities[11] of sound and the visible actions that produce it, hath made me entertain the following hypothesis.

[p. 6]

I suppose the air we breath[e] in to be a mixture of divers minute bodies which are of different sorts and sizes, though all of them are so small as to escape our senses. The grosser of them are elastical, and are resisted by solid bodies, which are in no sort pervious to them. The smaller parts pass through solid bodies, but not with that ease but that upon a sudden and violent start[12] of them, they shock[13] the parts of solid bodies that stand in their way, and also the

---

[10] *acquiesce from*: OED *v.* 1.b. *intrans. obs. rare* to rest, or cease from (1659).

[11] *qualities*, i.e., secondary, subjective properties of sound, phenomena; see also FN p. 28 (= stable or unstable relations of sounds in conjunction) and p. 29 (= pitch).

[12] *start*: OED *sb.*^{2} 5.d. act of setting in motion (1602–1904); see also starting, FN p. 6.

[13] *shock*: OED *v.*^{2} 1.c *trans. obs. rare* oppose (1667–76); hence, the action of elastic impact; i.e., *shock*: OED *sb.*^{3} 2. a sudden and violent blow, impact or collision tending to overthrow or

grosser parts of the air. And there may be another degree of most subtil ethereal parts with which the interstices of these and all other bodies are repleat, which find freer passage every where, and are capable of no compression, and consequently be the medium and cause of the immediate communication of [light to the organ of] sight. The middle sort of these I esteem the medium and cause of sound, and that at any time when the grosser air is driven off any space and leaves it to be possest by these and other more subtil bodies, and returns by its elasticity to its former place, then are these parts extruded with violence as from the center of that space, and communicate their motion as far as the sound is heard.

Or when any solid body is moved with a sudden and violent motion, these parts must be affected thereby. For as these parts are so much resisted by solid bodies as to shock them, so on the contrary they must needs be moved by the sudden starting of solid bodies.

So that sound may be caused by the tremble of solid bodies without the presence of gross air, and also by the restitution of gross air when it has been divided with any violence.

We see a bell will sound in the Torricellian space. And when the air is divided with any sudden force[—]as by the end of a whip having all the motion of the whip contracted in it, and by a sudden turn[14] throwing off the air; by accension,[15] as in thunder and guns; or by any impression of force carrying it where other air cannot so forcibly follow, as upon compressing of air in a bladder till it breaks, or in a potgun[16][—]a sudden crack[17] will be caused.

[p. 7]

I shall leave others to apply this hypothesis to the aforementioned[18] phænomena (which they may easily do) and proceed to the discourse of

---

to produce internal oscillation of a body subjected to it, also the disturbance of equilibrium or the internal oscillation resulting from this (1614–1726).

14 *turn*: OED *sb*. II. 8. change of direction (1412–1847).

15 *accension*: OED *sb. obs.* or *arch.* the action of kindling or the state of being kindled; ignition (1646–1801).

16 *potgun* = *popgun*, i.e., a child's toy, consisting of a short straight tube from the mouth of which a tight-fitting pellet is expelled with a pop by compressing the air in the tube with a piston. In a fragment of his writing preserved in *Life/FN*, p. 259, FN refers to compressed air in 'windgunns', so that, here, perhaps, he misremembered what he 'heard from others' about experiments made with air guns to show that air 'suffers itself to be crouded into a comparatively very little room'. For some of those experiments, see, e.g., Boyle, *The Works*, vol. 1, pp. 13, 14, 76, 140, 202; see also Webster, 'The Discovery of Boyle's Law', pp. 455, 458.

17 *crack*: OED *sb*. I. 1. a sudden sharp noise as of something breaking (1300–1867).

18 1677 text has 'afore mentioned'.

musick, where I am in its due place to shew how this action that causes sound is performed by the several instruments of musick.[19]

OF A TONE

A tone is the repetition of cracks or pulses in equal spaces of time so quick that the interstices or intervals are not perceptible to sense.

The more quick the pulses are, the more acute the tone is. Where the intervals are not equal, nor in musical proportion, the sound is not in tone[20] but an irregular noise.

The compass of musick extends from such tones whose intervals are so great that the several pulses are distinguishable by sense, to those [tones] whose interstices are so very small that they are not commensurate with any other.

In which compass the several tones are infinite in number as all space may be divided *in infinitum*. But the tones useful in musick are those within the scale, which are not very many, and they are placed in the scale as they have relation to one another.

OF THE RELATION OF TONES AND THE UNION OF MIXT SOUNDS

When the pulses of tones are coincident one with the other, there is an union of the sounds.

For when any sound is made, solid bodies within[21] the sphære of it are moved, and if the body moved stands upon a spring[22] (as all instruments producing tones do) it will vibrate by intervals according to the measure[23] of the spring until the force be spent.

---

[19] See FN pp. 11–16.

[20] *in tone*, i.e., produced by regular vibrations of a sounding source (OED does not give 'in tone').

[21] 1677 text has 'whithin'.

[22] *spring*: OED *sb.*[1] IV. 18. elastic energy or force *(a)* of the air (1660–1815), *(b)* of solids (1674–1879); see also FN pp. 11, 12, 14, 16, 17, 20.

[23] *measure*, i.e., comparison to some unit recognised as standard; see also FN pp. 12, 15, 16, 18, 20, 32, 34, 35.

Now if the moving force be repeated, as it must be where **[p. 8]** it proceeds from a tone, and the terms[24] of the force are equal to the terms of vibration, every new stroke strengthens the former impression, and after some continuance the vibration of a body affected by sound, may become so great as to be sensible to the eye and ear.

But where the terms of vibration are not commensurate with the terms of concussion, the following pulses destroy the motion begun by the former, and the vibration will never augment beyond what it was upon the first impression.

Upon this reason a string tuned in unison to another will manifestly shake when its fellow sounds, and another string that has a discording tone, though of the same bigness,[25] will not appear to be moved.

And for this reason, wheresoever [the terms of] the vibrations of two sounding strings are equal, the pulses will be coincident. For although one should set out after the other, yet by this affection [resonance] they would by the second or third pulse become coincident.

Where the terms are equal [i.e., 1 to 1], and the coincidence is at every pulse, it may be said [to be] the same sound though proceeding from several instruments. For the mixed sound has no alteration, but by being more loud and full by the addition.

Where the termes are 2 to 1, or 4 to 1, or 8 to 1, or 16 to 1 [as in *Figure 1.A*, Tab. i, 'Octaues'], there will be coincidence to every pulse of the base, and between those termes the treble is heard alone without any mixture, so that the sound is not changed but augmented by the addition of stronger pulses at proportionable intervalls, which makes the pulses superinduced appear of the same nature, though more grave or acute. This is the case of octaves.

Where the termes are as 2 to 3, every third pulse of the treble, and every other pulse of the base are coincident, but the intermediate pulses, namely, 2 of the treble, and 1 of the base[,] are not so, but keep certain distances, which makes the mixture produce an alteration. But the coincidences be- **[p. 9]** ing so frequent, and the distances of the intermediate pulses proportionable, the mixture is very pleasing, and is the principal chord called the fifth.

Where the termes of coincidence are further off, as when they are but the fourth, fifth or sixth pulses of the treble that are coincident with the third, fourth and fifth pulses of the base, yet the mixture is pleasing, and produces chords, which chords are more or less perfect, according as they are more or less coincident.

Coincidence upon every fourth pulse of the treble is the fourth.

---

[24] *terms*, i.e., periods of time; hence, *(a)* frequencies—the number of complete vibrations executed within a determinate period of time (e.g., a second); *(b)* each of the two quantities composing a ratio, or a fraction, that represents such frequencies; see also FN pp. 9, 11, 13, 29.

[25] *bigness*, i.e., length; see also FN pp. 11, 12, and cross-section, see FN p. 11.

Every fifth pulse is the third sharp.

Every sixth pulse is the third flat.

When the fifth pulse of the treble goes with the third of the base, it is a sixth sharp. The eighth pulse of the treble going with the fifth of the base makes a sixth flat.

That these chords have their pulses in such proportionable time is demonstrable by the division of the monochord, which being stopt, in 1/2 gives an octave, in 1/3 gives a fifth, in 1/4 gives a fourth, in 1/5 gives a third sharp, in 1/6 gives a third flat, [in] 2/5 gives a sixth sharp, [in] 3/8 gives a sixth flat. This is agreed by all that ever treated of musick, therefore I shall speak no more of it.

It may seem strange that a coincidence in this manner should unite sounds, and that the five and four pulses that are disagreeing should not hinder more, than the sixth coincident pulse could reconcile; and that a coincidence between the eighth pulse of the treble with the fifth of the base should make the sounds agreeable, notwithstanding the variance of those which are intermediate.

But this will be very clear, when it is observed that the intermediate pulses do not at all hinder, for they are all placed in such manner, in relation to one another, that where any of them distract[26] the pulses of the concording string on the one hand, there are others that[,] by being just as much on the other hand, set them right again. They are in balance **[p. 10]** one against another, so that it is as if there were no intermediate pulse[s] at all, as to any effect of hindring the coincidence.

And the intermediate pulses being at such symmetrical distances[,] a new proportion arises from the mixture of them, which makes the joined sounds pleasing to the sense.

Where the coincidences are more remote, and all the intermediate pulses disagreeing, as 7 to 6, 8 to 7, 9 to 8, 10 to 9, the by pulses are too many to be kept steddy, and the pulses on [the] one hand prevail too far before those on the other hand come to set them right, so that the sounds will not maintain their coincidences, but produce jarring and discord.[27]

When sounds are in discord, the concussions caused by them are in opposition one to another; and the organ of sense cannot be affected by both together, but they will appear ragged and jarring, and extream noisome. Nay some sounds will be so contrary to one another, and so hard a match, that they cannot sound both together, but each hindering the other will take turnes. Thus do they [organ makers] imitate the drum in an organ, by joyning two deep

---

[26] *distract*: OED *vb.* 1. *trans. obs.* to draw in different directions (1600–1651).

[27] *discord*, i.e., the effect due to beats—'jarring' or intermittent tones—produced by pitches of close but not identical frequencies; see also FN pp. 15, 17, 20, 28.

discording notes, that rumble in that manner sounding by turns.[28] And in tuning organ pipes it may be observed, that the sounds of two pipes whilst in discord will flutter,[29] but when the discording pipe is brought to the other's tone, the sounds will as it were jump into tune, as if they had attracted one another, and afterwards one clear and entire sound will be produced.[30]

Upon this relation of tones depends the scale of musick, of which I shall shew the degrees, and the reason why they are so placed, when I come to speak of schismes.[31]

Before I leave this head I must observe, that octaves not altering the manner of sound, but giving a fulness in duple proportions, which are comprehended in the same time, it follows, that all octaves are easily brought in mind, and are as it were included in any tone.

This being considered, it will appear, that there can be no other variety of coincidence within the compass of seven pulses that is not allowed in musick, and produces some chord. For [as in *Figure 1.A*],

[p. 11]

1/3 is as 2/3[,] 2 being the octave to 1.

1/5 is as 4/5[,] 4 being in octave to 1.

2/5 as 4/5, 2/6 as 2/3, 1/6 as 4/6, as [is] 2/3.

And the eighth coincident pulse in a sixth flat, comes in upon this contemplation, but as the reverse of a third sharp, 5/8 being the reverse of 5/4, and the other sixth 3/5 is the reverse of 6/5. So that there can be no other coincidences under the distance of the seventh pulse, but what have the names of chords, or the octaves; and we are not to wonder why there are no more concording notes.

HOW TONES ARE PRODUCED, AND OF ASSISTANCES TO THE SOUND BY INSTRUMENTS

Wherever a body stands upon a spring that vibrates in equal termes, such a body put into motion will produce a tone, which will be more grave or acute, according to the velocity of the returns.

Therefore strings vibrating have a tone according to the bigness or tension of them; and bells that vibrate by cross ovals, produce notes according

---

[28] *by turns*: OED *sb.* V. 28.b. successively (1538–1850); see also FN p. 13.

[29] *flutter*, i.e., beat (like a drum); see also fluttering, FN p. 15.

[30] That imperfect octaves and fifths produce beats was long known to organ makers, who made use of it, in practice, to obtain the required just or tempered intonation; see, Helmholtz, *On the Sensations of Tone*, p. 181.

[31] See FN pp. 24–28.

to the bigness of them, or the thickness of their sides. And so do all other bodies whose superficies being displaced by force, result[32] by a spring which carries it beyond its first station.[33]

It is easie to comprehend how every pulse upon such vibrations causes sound. For the gross air is thrown off by the violence of the motion, which continues some moment of time after the return of the vibrating body. Whereupon some space must be left to that subtil matter, which[,] upon the result[34] of the air[,] starts[35] as from a centre; which action[,] being the same I supposed to be the cause of sound,[36] is repeated upon every vibration.

It is more difficult to shew how tones are made by a pipe, where there are no visible vibrations. I will consider the frame of a pipe, and the motion of the air in it, and there- **[p. 12]** by attempt to find the cause of the tone of a pipe, and the pulse that gives the sound.

There is in any pipe a cavity of a certain gage[37] or bigness which is exposed to the outward air, but in some parts of it [air is included], and according to that cavity is the tone of the pipe.

The air in any cavity being capable of being compressed by outward force, if the cavity be large, [the air] easily yields to a compression; but if it [the cavity] be small, [the air] is more hardly compressed. When the compressing force ceases, the body that was compressed with difficulty flies out with sudden violence; that which is more easily compressed restores it self not so soon. Whereby it happens, that every cavity has a certain tone according to the measure of the vibrations of the air in it, which it will keep what way soever the force comes to it, as the water in a shallow or creek moved by the large waves of the sea, retains not a motion in such huge waves, but in waves proportionable to its own bigness and depth.

Hereupon I conclude, that any pipe, whether made of wood, stone, glass or metal, has a certain and determinate tone, according to its gage and cavity, which is the natural tone which will be produced by blowing it. If the cavity be small, the sound will be acute, and may be made more acute by apertures, whereby the cavity is diminished, and the inclosed air exposed to the outward air in greater measure.

---

[32] 1677 text has 'results' = *result*: OED *v.* 2.c. *obs.* recoil, rebound, spring back (1598–1784).

[33] *its first station*, i.e., its equilibrium position; see also its first place, FN p. 18.

[34] *result*: OED *sb.* 1. *obs. rare* the action of springing back again to a former position or place (1626).

[35] *starts*: OED *v.* I. *intrans.* 2.a. moves with a violent impulse (c.1290–1859).

[36] See FN p. 6.

[37] *gage* = *gauge*, i.e. linear measurement in a particular direction, as length, breadth, thickness; see also FN pp. 12, 13, 19.

When I say, the tone is according to the proportion of the cavity, let the pipe be made of what it will;[38] I do not deny but that if the materials of the pipe be soft and yielding, it may give some small alteration to the tone from what it would be if they [the materials] were of a substance hard and firm, where the impelling air would find a more brisk resistance. But the difference is inconsiderable, and the spring of the air depending mainly upon the greatness of the cavity, I take no notice of it.

To shew how the pulses are caused, whereby the included air is put into this motion, it is necessary to observe the **[p. 13]** frame of a pipe, which chiefly consists in having a long slit through which the air is blown in a thin film against or very near a solid edge that is at some distance opposite to it, in such manner that the intermediate space is covered by the stream of air.[39] This film of air on the one side is exposed to the outward air, and on the inside is defended from it by the sides of the pipe, within which the air inclosed in the pipe stagnates, whilst the outward air is by the blast put into a vortical motion.

The vortical motion or eddy on the outside is so strong, that there not being a balance to that force on the inside, the film of air gives way, and the eddy bears into the pipe, but is immediately overcome by the blast, which prevails untill the eddy overcomes it again. And so there is a crossing of streams by turns and pulses, which causes the voice of the pipe, the gross air of one stream being thrown off by the interposition of the other.

These vicissitudes[40] or termes will answer the tone of the pipe according to the gage of its cavity. For the spring of the included air helps toward the restitution of the blast and eddy in their turns, which causes those turns to comply with the tone of the pipe. And therefore the same blast will cause several tones, if the gage or measure of the included air be changed by apertures in the side of the pipe.

But there must be some proportion between the mouth, (so I call that part of the pipe where the voice is) and the gage of the pipe. For though the pulses will be brought to comply with the tone of the pipe in any reasonable degree, yet when there is great disparity they[41] [the pulses] will not do so; as if the pipe be too long for the proportion of the diameter, the pulses at the mouth cannot be brought to so slow termes as to answer the vibrations of the included air.

---

[38] 1677 text has 'well'.

[39] FN here describes a flue pipe, which is driven by edge tones; see Chapter 2, p. 59 *supra*.

[40] *vicissitudes*: OED *sb*. 6. *obs*. reciprocation, return (1565), *pl.* = complete vibrations (going *and* returning); see also FN p. 14.

[41] 1677 text has 'it'.

Therefore the pipe will not speak unless it [the included air] can break[42] into some higher note. If the filmy stream of air be too thick, the pipe will not speak, because the eddy cannot break through. If the opposite edge be too remote, the stream cannot entirely cover **[p. 14]** the aperture, for it mixes with the outward air, and is more confused the farther it is from the vent or passage, whereby some outward air may have communication to make an opposite eddy on the inside of the stream. For the same reason, if there be the least aperture in the region of the mouth of the pipe, it will not speak at all.

Hence is it, that the voice of organ pipes is so tender and nice. But shrill whistles depend not upon this ground; for they are made in any small cavity where the blast is so applyed, that the erumpent[43] air must cross it, whether the stream be thick or thin. Therefore the bore of a key, a piece of nutshell, or any other cavity will make a whistle, whose tone will be according to the quantity of the included air, for the less that is, the harder it is to be compressed, and the quicker and stronger it must break forth.

Another kind of whistle[44] is, when a hollow body with a small cavity is perforated by opposite holes, a blast either way will cause a tone, which seems to be made in this manner.

The air that is violently drawn or thrust through these holes is straitned at the passage by the swiftness of the motion, and within the cavity [the air] is somewhat enlarged, and consequently its force is directed, and it presses beyond the compass of the opposite aperture, whereupon it bears on[45] all sides into the cavity. Hereby the air within the cavity is compressed untill it breaks forth by crossing the stream, which[,] being done by vicissitudes[,] causes a tone. This kind of action, as I imagine, is performed when men whistle with their lips.

In some pipes the pulses are caused by springs,[46] as the regoll[47] stop of an organ, which is commonly tuned by shortning the spring whereby it [the sound] becomes stronger, but the note will be changed by the alteration of the

---

42 *break*, i.e., divide into separate vibrating segments, thereby varying the pitch; see also FN pp. 15, 17–21.

43 *erumpent*: OED *a.* that bursts forth (1650–1871).

44 *another kind of whistle*, i.e., either a flageolet or a recorder (in England, the latter superceded the flageolet from about 1670).

45 1677 text has 'of'.

46 *springs*, i.e., resilient tongues (acoustic vibrators); see also spring, FN pp. 14, 15.

47 *regoll*, i.e, regal; one of a family of organ stops.

cavity. And therefore to make them steddy, some that stand upon very weak springs have pavelions[48] set to them.

A rustick instance may be given of the compliance of a **[p. 15]** spring, in taking such vibrations as are proportionable to the cavity. It is a jews-harp, or jews-trump,[49] the tongue whereof has natural vibrations according to strength and length of the spring, and so is fitted to one particular tone. But countrey-men framing their breath and their mouth to several notes make a shift to express a tune by it.

In a shawm or hautboys the quill[50] at the mouth is a kind of spring, but so weak and indifferent that it complies with any measure, and therefore the tone will be according to the apertures of the pipe.

The fluttering and jarring of discording sounds, which I before did observe,[51] is regular, and the sounds take their turns with equal interstices, which makes the joyning of them produce a harsher sound than either had before. Whereby organ-makers imitate the hautboys or trumpet,[52] without any spring or quill, by joyning discording pipes.

In a sackbut[53] the lips of a man do the same office as a quill does in a shaume or hautboys; when the included air is lengthned, the tone varies. Nevertheless they [performers] can produce several notes that are in chord to

---

[48] *pavelions*: (French: pavillion; English: pavillon) bell of a wind instrument. Here, a reference to a type of regal stop—possibly the Trichter-Regal, a funnel-shaped or conical resonator that looks like a small trumpet; see Wedgwood, *A Comprehensive Dictionary of Organ Stops*, pp. 132–135; or possibly a bass viol unison stop, which, in England, was made conical with a bell top; see Alexander Ellis in Helmholtz, *On the Sensations of Tone*, p. 93.

[49] *jews-harp, or jews-trump*, i.e., an instrument with the same natural series of tones as the trumpet. It consists of a frame of wood or metal in horseshoe form, to which one end of a flexible lamella is attached, its other end remaining free. The narrow end of the frame, held loosely between the teeth so that the lamella can swing freely, is plucked with a finger, causing it to vibrate in the player's mouth, which thereby becomes a resonator. Depending on the position of tongue, lips and cheeks, one or another tone in its natural series can be emphasised.

[50] *quill*: OED *sb.*[1] 1. *obs.* a hollow stem or stalk, as that of a reed (1412–1688). Here, the mouthed double reed (acoustic vibrator) of the shawm and oboe ('hautboy'). But the reed seems to have been attached to the quill of a feather, since, in ff. 74–74v of 'The Theory of Sounds...', (GB:Lbl) Add MS 32534: ff. 1–82v, RN stated that

> the materiall [of reed or cane] is shaved very thin and put together as 2 plates, the one end (almost) closing flatt, which is for the lipps. The other [end is] lapped and made fast about a quill, which is to be inserted at the end of the tube, or pipe [of the instrument]. The reed of it self thus disposed is an instrument of sound, but very shrill ... When the reed is inserted in the tube or pipe, then the body or columne of air included, as so much of weight is added to the air of the quill, ... takes downe the tone to a flattness accordingly.

[51] See FN p. 10.

[52] But see FN p. 10, where he stated that organ makers utilise beats to imitate a drum.

[53] *sackbut*, i.e., early name for the trombone.

the tone of the instrument, by strengthning the blast without lengthning the cavity. And in a trumpet,[54] which is the same kind of instrument, (only not capable of being lengthned) they can sound a whole tune, which is by the artificial ordering the blast at the mouth, whereby the sound breaks into such notes as are to be used.[55]

This depends upon the consideration of the varying and breaking of tones into other notes, whereof I am to speak particularly in my next chapter, to which I hasten.[56]

Having shewed how tones are produced by instruments of musick,[57] I must take notice of other assistances instruments give to the sound.

In violins and harpsechords the tones are made wholly by the vibrating strings, but the frame of the instrument adds much to the sound. For such strings vibrating upon a flat rough board would yield but a faint and pitifull sound.

[p. 16]

The help that instruments give to the sound, is by reason that their sides tremble and comply with [the terms of vibration of] any sound, and strike the air in the same measure that the vibrations of the musick are, and so considerably increase the sound.

This trembling is chiefly occasioned by the continuity of the sides of the instrument with the vibrating string. Therefore if the bridge of a violin be loaded with lead, the sound will be damp[ed]; and if there be not a stick called the sound-post to promote the continuity between the back and belly[58] of the instrument, the sound will not be brisk and sprightly.

Such a continuity to the nerve of hearing will cause a sense of sound to a man that hath stopped his ears, if he will hold a stick that touches the sounding instrument between his teeth.[59]

The sound of it self without such continuity would occasion some trembling, as may be seen by the moving the unison strings in the instance before given.[60] But this is not considerable in respect of the other [instance],

---

54 *trumpet*, i.e., the natural or valveless instrument, most typically with a fundamental of either eight-foot C or the D of slightly over seven feet in length.

55 The process described in this paragraph is called ‘overblowing’, whereby an air column breaks up into aliquot parts, each of which will vibrate at a frequency that is in direct proportion to the fundamental frequency; see also FN p. 19.

56 See FN pp. 17–21.

57 See FN pp. 11–15.

58 *belly*, i.e., the front bulging surface of the violin.

59 For hearing by bone conduction, see Chapter 2, pp. 57–58 and n.70 *supra*.

60 See FN p. 8.

though it be all the assistance that the structure of a chamber can give to musick, except what is by way of eccho.[61]

This tremble of the instrument changes with every new sound. The spring of the sides of the instrument standing indifferent to take any measure, receives a new impression. But a vibrating string can take no measure but according to its tension.

Therefore instruments that have nothing to stop[62] the sounding strings make an intolerable jangle to one that stands near, as bells to one that is in the steeple, and hears the continuing sound of dissonant tones; such is the dulcimer.[63] But the harpsechord that hath raggs[64] upon the jacks by which the vibration of the string is staid, gives no disturbance by the sonorousness of the instrument, for that continues not the sound after the vibrations [are] determined, and another tone [is] struck, but changes and complies with [the terms of] the new sound.

[p. 17]

OF THE VARYING AND BREAKING OF TONES

Tones will by accident vary and break into other tones,[65] which are chords to the natural tone of the instrument, which in pipes is so regular, that the compass of the musick depends upon it. Upon strings it [breaking] will happen, but not so gratefull to the ear, nor much in use. I will endeavour to explain how it is caused in strings, and speak of pipes afterwards.[66]

A perfect string produces a clear sound by entire and equall [timed] vibrations, there being no inequality to hinder the motion from being uniform from one end to the other, according to the laws of a pendulum. But if the string hath any inequality towards one end, it will yield a jarring and

---

[61] FN alludes briefly to room (architectural) acoustics, which, as a field of investigation, dates from the end of the nineteenth century, when Wallace Sabine began a systematic study of three problems—loudness, distortion of complex sounds by interference and resonance, confusion by echo, reverberation and extraneous sounds; see Beyer, *Sounds of our Time*, pp. 186–187.

[62] *stop*, i.e., damp by means of a vibration damper, a device that either reduces the amplitude of the vibration or reduces the communication of vibration energy from the vibrator to its surrounds.

[63] *dulcimer*, i.e., a struck zither. Pepys, *The Diary*, vol. 3, pp. 90, 118, recorded hearing one for the first time in 1662.

[64] *raggs*, i.e., thick cloth such as felt that constitutes a damper.

[65] *other tones*, i.e., secondary or subordinate tones produced by the vibration of the aliquot parts of a sonorous source, usually accompanying the fundamental tone produced by the vibrations of the source as a whole.

[66] See FN p. 19.

distracted[67] sound. For the resistances are not only at the ends of the string, but there are cross tugges that alter the course of the vibrations; which is evident in the manner whereby musicians try if their strings be true. For if the string be true[,] the vibrations will appear as a clear filme; but they will appear with cross threads if the string be false.

If there be a long string pendulum, the motion will be uniform and synchronous[68] from the bottom to the top. But if there be a weight fastned to the string towards the top, it will hinder the regularity of the motion. For the upper weight will be put into motion, and will have vibrations of its own, which will be mixed and confounded with the vibrations of the lower weight. In the same manner in a false string two or more sorts of vibrations are blended, which are not in proportion, but casual and discording, and therefore instead of a tone produce an horrible jarr.

Now as a string by such small accidents is brought to vibrations of several parts of the string that hold no proportion one to another, much more easily may it be imagined to have vibrations of uniform parts, that shall be syn-**[p. 18]** chronous, and produce one single sound; as if the parts that vibrate severally are half parts, seconds or thirds of the string. And how this may be caused, I will enquire.

If a string pendulum be very long, and the weight at bottom be not sufficient to cause the whole string to divide the air; at first the pendulum will be sustained by the resistance of the air, and afterwards it will put the whole string into motion: but not that of a streight line. For the pendulum will return before the direct motion performed at the top.

This will occasion thwarting of motions, which being balanced between the force at the bottom and the check at the top, the crossing point will settle in such a division of the string, that they [the motions] will continue uniform till the force be spent.

I imagine, that if a musical string be so struck, that the whole string is not removed out of its place to cause the greatest vibration in the middle, there must be a crossing of vibrations. For before the motion [is] communicated to the farther end, the part of the string that was struck may have restored it self to its first place. This crossing of vibrations, if the string be true, will be upon equall parts, and produce an octave, fifth, or some other chord.

It is common experience, that a great string struck near the bridge with a bow where the rosin takes but small hold, will whistle[69] and break into chords

---

[67] *distracted*: OED *ppl.a.* 2. *obs.* exc. as *fig.* agitated, disturbed (1632–1845); here, beating.

[68] *synchronous*: OED *a.* 1. happening at the same time; coincident in time; belonging to the same period, or occuring at the same moment, of time, simultaneous (1669–1878); see also FN pp. 17–18 and, for a different meaning, p. 20 n.81.

[69] *whistle*, i.e., produce flageolet tones (flute-like sounds—‘shrill notes’).

above; which if it were struck by the thumb that removes it out of its place, would give the true tone.

The trumpet marine that sounds wholly upon such breaks, is a large and long monochord play'd on by a bow near the end, which causes the string to break into shrill notes. The removing the thumb that stops upon the string gives measure to these breaks, and consequently directs the tone to be produced. The jarr at the bridge takes the same measure and makes the sound loud, in imitation of a trumpet, which otherwise would be like a whistle or pipe.

[p. 19]

The touch of the thumb[70] less hinders the sound of the string when it is upon the point[71] where the vibrations cross, than when it is in any other part. For we see when any great string has an entire vibration, such a touch would immediately extinguish the sound. This makes the trumpet marine, with the thumb placed upon it, take to such a note, as that the division of vibrations shall lye just under the thumb.

Speaking of the counterfeit trumpet, I must observe that the true one seems to give all its notes by way of breaking, which causing the metal to jarr gives so loud a sound. I fancy the true gage of the air included in a trumpet, in comparison to[72] its bore[,] would fit an eight[h] below the note they [performers] call the note of the trumpet, because a fourth below it is so easily sounded, which is a fifth to the true tone, in case it be placed eight notes lower. As it easily divides into three to give a fifth, so it may well treble that division, and give a second to the trumpet note.[73] And for the rest of the notes which a trumpet will easily produce, they are the third sharp, fourth, fifth and sixth sharp, which arise upon the most easie divisions of the monochord [the trumpet marine], and therefore [are] most readily produced by breaking, when the strength of the blast and the action of the lips direct it. The trumpet not being able to answer a flat key, the sackbut[,] as I observed before[,][74] is made in

---

[70] 1677 text has 'thumb-less'.

[71] *point* = *node*: OED *sb.* 6.a. a point or line of absolute or comparative rest in a vibrating body (1831–79).

[72] 1677 text has 'of'.

[73] A natural series of overblown tones with C as the fundamental will produce D by overblowing G, the third tone in the natural series. The second note in the example below is the performer's 'note of the trumpet':

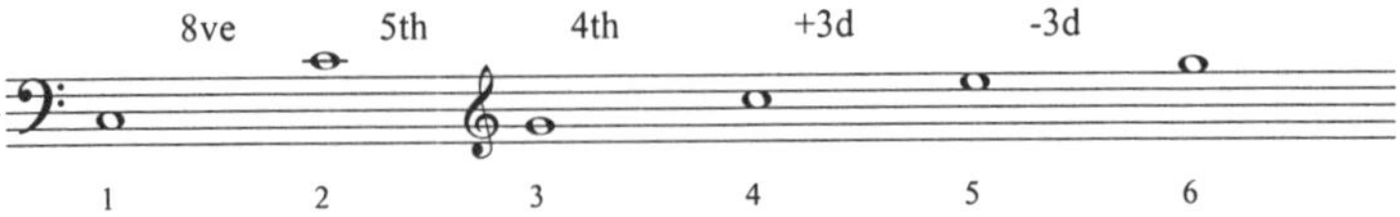

[74] See FN p. 11.

such manner that it may be lengthned to supply this defect, whereby it will give any note at the pleasure of him that plays.[75] But when he is to sound an eighth or fifth above, he never lengthens the instrument, for without any remove of his hand, the note easily breaks into those higher chords.

The trumpet is of the nature of pipes, and therefore properly leads me to speak of the varying notes upon pipes, which is so regular, and so much in use, that the compass of the musick depends upon it.

I have shewed that sound does cause a motion not only of solid bodies, but of the grosser parts of air within the sphere of it.[76]

[p. 20]

Now if the air that is moved by being inclosed stands upon such a degree of resistance to compression, that it has a spring vibrating in the same measure with [that of] the sound that puts it into motion, there will be the same [resonance] effect as when two strings are tuned in unison, that is, the motion will be so augmented by succeeding regular pulses, that the enclosed air may be brought to ring and produce a tone.

Vitruvius in his *Architecture* advises, that in making a theatre there should be vases or hollow pots of several sizes to answer all the notes of musick placed upon the stage in such manner, that the voice of them which sang upon the stage might be augmented by the ringing of them.[77] And he mentions divers ancient theatres where such were, in some of brass, in some of earth.[78]

An organ-pipe of a very deep base will not speak suddenly when it is alone; but if an octave be in play at the same time, it will answer the quickest

---

[75] In *Cursory Notes of Musicke*, p. 119, RN reported that

> trumpetters have found some expedients, as the sackbut, which prolongs the tube ad libitum. But a more ingenious devise is that which Mr. Shoar [Shore], a most exquisite trumpetter, used to adapt his trumpet to consort. It was by an imperceptible lengthning or shortning the tube, by the help of a screw or worme, so as by turning the trumpet round he gained, or remitted of the length, by which his exotick notes fall all into use.

There were three trumpeters, all related, by the name of Shore, each of whom had court appointments; see Ashbee and Lasocki, *A Biographical Dictionary of English Court Musicians*, vol. 2, pp. 1003–1008. RN probably refers to John Shore, the trumpeter who has been credited with the invention of the tuning fork.

[76] See FN p. 6.

[77] Vitruvius Pollio, Roman engineer and author of *De architectura* in ten books, the fifth of which is thought to contain the first reference to acoustic vases along with an elucidation of how such a system of vases might be used in a large theatre. The only surface for the reinforcement of sound by reflections was the wall behind the stage, so Vitruvius suggested selective resonators, i.e., cavities of air that, by their resonant response to a voice suitably pitched, might reinforce the sound by resonance. For details, see Hunt, *Origins in Acoustics*, pp. 34–35, and Green, *The Harmonic Series*, pp. 107–109.

[78] *earth,* i.e., clay.

touch, because the tone of the lesser pipe being in chord causes a motion of the air included in the greater pipe, and the vibrations of the included air being coincident with those of the tone of the pipe, promote its speaking by putting the pulses at the mouth into a regular motion, or at le[a]st assisting thereto.

These instances shew the manifest effects of sound. And certainly sounds in discord are in contrariety [i.e., beat], and offer violence one to another; but sounds in musical chords are consistent,[79] and united together, and promote each other, as I have before shewn in what I said concerning the relation of tones.[80]

The synchronous[81] motion of the pulses at the mouth of the pipe with the vibrations of the included air promote the sound of the pipe, and it is hard for a pipe to speak where those motions are in opposition. But if by strengthning the blast[,] the pulses at the mouth are so quickned that they bear a musical proportion to the vibration of the air enclosed, as 2 to 1, 3 to 2, 4 to 3, 5 to 4, etc., there will be no hindrance but some advantage; so that it may be easily imagi- **[p. 21]** ned that the pulses at the mouth in such case will receive no interruption by the included air, but may maintain their stations, and produce a tone varying in the same proportion from the tone of the pipe.

It may be objected, that if it were so, there would be a mixture of sounds; and not only the tone that is in chord to the tone of the pipe would be heard, but also the tone of the pipe with it.

To which it may be answered, that the vibration of the included air is not so considerable as to be heard distinctly, though it may give alteration to a sound; as we see,[82] when a pipe gives the true tone, it speaks more in the belly[83] than when it breaks into an octave. But to wa[i]ve that answer, perhaps when the note breaks, the column of included air divides into proportionable parts, by reason that the vibrating force at the mouth, and the resistance at the aperture[,] meet one another in the middle, or upon other divisions, which will make the vibrations of the included air coincident with every pulse at the mouth of the pipe. Which is the more probable, because a very small aperture in the middle of a pipe will make it break into an eighth above, which seems not to happen by any diminution of the included air, but [by] interrupting the intireness of the column.

---

[79] *consistent*: OED A. *adj.* 6.b. *absol.* of two or more things, in agreement (1651–1884); see also FN p. 21.

[80] See FN pp. 7–11.

[81] *synchronous*: OED *a.* 2. recurring at the same successive instants of time; keeping time with; going on at the same rate and exactly together; having coincident periods, as two sets of vibrations or the like (1677–1899). FN's statement is cited as the earliest example of this use.

[82] 1677 text has 'see, that'.

[83] *belly*, i.e., interior (of the concavity).

OF THE NATURE OF KEYS IN MUSICK, AND OF A SINGLE TUNE

Before any note [is] struck[,] the ear is indifferent to all sounds, but when any tone is heard, that indifferency is determined, and the mind [is] taken up with the present sound. If the second note be not of relation to the first, the impression made by the first will be wholly obliterated and forgotten, the sounds will be disjoynted and incoherent. But if the second note be of relation to the first, it is consistent **[p. 22]** with the memory of the first. If the third note be more in chord to the first, then is that remembrance continued and strengthned. If succeeding notes are in discord, or of less relation to the key note, the mind and fancy are[84] drawn from the key note by the present sound. But by repeating the key note and its relatives with loudness and emphasis, the memory of the key note may be restored again.

Thus in passing to and from the key, by notes that have relation to the preceding notes[,] may a tune proceed to any length, and amuse the hearer. But when the fifth, which is the principal chord to the key, and most strengthens the memory of it, comes in play, nothing can be more perfect and satisfie expectation but the key note it self, which gives a perfect acquiescence[85] to the hearer, and a conclusion to the tune.

For this reason all tunes end in the key note, and the next preceding note is the fifth above.

In the progress of a long tune[,] other notes may gain so great impression in the mind by [its] dwelling upon the relatives to them, that the first key may be forgotten, and [in] the same way new notes may gain possession, so that the key may change many times. Thus it is in long fancies and preludes, and it is no matter in such case what key the tune concludes in. But in short airy[86] tunes there are but small digressions, leaning to the fifth or third, or other relative notes, and sometimes suddenly closing upon them, which does not destroy the memory of the first key note. And therefore the ear is not satisfied with such close, but expects another strain, which ending in the key note concludes the song.

Generally speaking, a tune must begin and conclude in the key note, because that note takes possession of the ear. But this is not such a rule but that sometimes a tune begins or ends upon notes relative to the key; and the key is shown by them, though it be not struck, as in the progress of a tune[,] the key is to be traced by the relatives to it, **[p. 23]** as well as by the [key] note it self.

---

84 1677 text has 'is'.

85 *acquiescence*: OED *sb.* 1. rest, quiet satisfaction (c.1631–1867); see also FN p. 29.

86 *airy*: OED *a.* II. 6.c. lively, sprightly (1644–1826); see also FN p. 31.

Hereby it appears that any note may be the key note, as to the pitch; [and] as to the species or sorts of keys they are two, viz., flat and sharp.

When a tune begins with a flat third it must continue so. For the third being a relative to the key note, is joyned with the memory of it, and become[s] as it were part of it, and a third sharp being discord to it would disjoynt the tune, though it be a better chord to the key note; it being necessary that the tune should proceed with regard to the preceding notes. For the same reason a third sharp when once taken into play must be continued.

When the third is flat or sharp the sixth must be so likewise, else the third to the key[,] which is of principal regard and relation to the key, and sometimes hath closes upon it, will have neither a fourth above, nor a fifth below in the scale, but a false fifth, which could not set it off in that lustre it ought to have.

There is no natural difference betwixt the musick in one key note or another; accidental differences there are. For in some keys the emphatical notes lying high are more sprightly than in others. Upon some instruments, by reason of holdings,[87] or open strings[,] the emphatical notes are of a mellow and durable sound, which gives advantage to the key. Upon most instruments the emphatical notes are not in tune, by reason of the schismes[88] of musick, (whereof I am presently to speak[89]) therefore such keys are not in use, and when play'd upon yield but a harsh tune. But these differences are but accidental, and relating to the advantages or imperfections of instruments.

I shall collect from what hath been said, that the musick of a single tune consists in the succeeding notes having a due relation to the preceding, and carrying their proper emphasis by length, loudness and repetition, with variety that may be agreeable to the hearer. This is called the formality of a single part, wherein musick is so copious, that it has afforded to every age and nation a several fashion, and to every musician a particular stile.[90]

[p. 24]

Graces in musick depend upon the same reason with the formality of a single tune, and skilfully applyed do much advance it, by strengthning the memory of preceding, and the expectation of succeeding notes, the doing whereof is left to every player.

---

[87] *holdings*, i.e., perhaps, notes that 'bear long continuance, with emphatick dwelling upon them'; see f. 2v of RN's 'Some Memorandums, concerning Musick', (GB:Lbl) Add MS 32532: ff. 1–26v.

[88] *schismes*, i.e., variations in frequency of two nearly identical pitches that arise during the process of changing from one tonality ('key') to another, so that some kind of temperament is necessary; see also FN pp. 24, 27.

[89] See FN pp. 24–28.

[90] See also FN p. 35.

OF SCHISMES, AND THE SCALE OF MUSICK

There are schismes in musick, because the scale is naturally fitted to one key; and though it will indifferently serve to some others, yet it will not to all.

The scale of musick is not set out by any determinate quantities of whole notes or half notes,[91] though the degrees are commonly so called. But the degrees in the scale of musick are fixed by the ear in those places, where the pulses of the tones are coincident, without any regard to the quantity. I shall endeavour to show how all the notes come into the scale by their relation and dignities; and then it will be obvious why[,] for easiness of instruction and convenience[,] the scale of degrees of musick is made as the musicians now exhibit it.

The key note being given, the first note in dignity is the fifth above, which must have its true place. For whilst the tune is in the key, it has the greatest emphasis; and upon any digression from the key, this note is nearest to usurp it. And therefore when the chords to the key note are placed, the chords to this note are to be [re]garded,[92] for without them there cannot be any variety, this being the first step from the key.

The other chords to the key are the third flat and sharp, the fourth and the sixth flat and sharp.

The chords to the fifth above the key are the 2^{d} and 7th flat and sharp, which are the fifth and third flat and sharp to that note.

[p. 25]

When the fifth has usurped the key, the fifth above that must have the same emphasis, and will bring in the sharp fourth, or false fifth, which is its third sharp.

The half note above the key is farthest from the key's relatives, and therefore the sound of it removes the dominion of the key note farther off than any other. The nearest way of bringing it in request is when the fifth below has usurped the key, to which it is a sixth flat. The fifth below taking the key makes an ingratefull remove, for it is not a chord comprehended in the key note, but the key note is comprehended in that, and so it diminishes the regard to the key note more than any other, making the transition to the key note, not as a restitution, but as a digression to a secondary key, which should not continue.

These are all the notes of the scale within the compass of an octave, and by this method should they be tuned, viz., third flat and third sharp, fourth, fifth, sixth flat and sixth sharp, by the key note[;] second, and seventh flat[,] and seventh sharp[,] by the fifth above the key or its octave; false fifth or fourth

---

91 *whole notes or half notes*, i.e., steps or degrees in the scale; see also FN p. 25.

92 1677 text has 'begarded'.

sharp must be third sharp to the second[;] and the half note above the key must be a sixth flat to the fifth below the key.

An instrument thus tuned will be exact to that one key, and serve to play tunes in that key. But when the musick turns to other keys that are not relative to it, the divisions and degrees that fitted the first key will not sute with the others, but must be removed, and thereby the schismes appear, without which the notes which ought to be principal and most perfect chords to the new key, will be manifestly out of tune.

This will best appear upon view of the division of the monochord, which I have set down in the plainest manner I could frame. **[p. 26**[93] see *Figure 1.C***]** It must be observed that the monochords are not divided as they ought to be; the reader therefore is to regard only the figures, and not the length of the lines. **[p. 27]**

By these divisions it appears, that the tone above the key, and that between the fourth and fifth, and that between the sixth and seventh, whether flat or sharp, are ninths of the string, which is called the tone major; the other tones are tenths of the string, and the natural semitones are sixteenths.

Other semitones which are placed in the scale, by the rules before mentioned[,][94] have other proportions,[95] because neither tone major nor tone minor can be divided into sixteenths. Therefore if one semitone be a sixteenth in a tone major, the other will be 19 2/7, and if one semitone be a sixteenth in a tone minor, the other will be a 25th, which is the semitone[96] that naturally lies between a third flat and a third sharp. If a tone minor be so divided, that one semitone is 25th of the string, the other will be 13 1/2.

These unequall tones and semitones having their proper places in relation to the key note must needs produce schismes upon displacing the key note. I shall give one instance. Suppose the key note a tone lower, which is at 400 upon the monochord in the first scheme [see *Figure 1.C*, upper part], then by the present scale the fifth should fall upon the fourth to the former key, which is 270. But by the division of the monochord[,] 400/3 produces 133 1/3, whose complement is 266 2/3; so that the schisme there is three and 1/3, which makes a vast difference, and is occasioned, because, whereas there should be two tones major, one tone minor, and a semitone to make a perfect fifth, this consists of two tones minor, one tone major, and a semitone.

It will be easie, by turning the octave line divided into semitones into a circle, and supposing the key note upon other semitones, to find the differences

---

93 1677 text is unpaginated; see also *Figure 1.D*, pp. 167–170 *infra* and p. 133 *supra*.

94 See FN pp. 24–25.

95 1677 text has ‘porportions’.

96 1677 text has ‘somitone’.

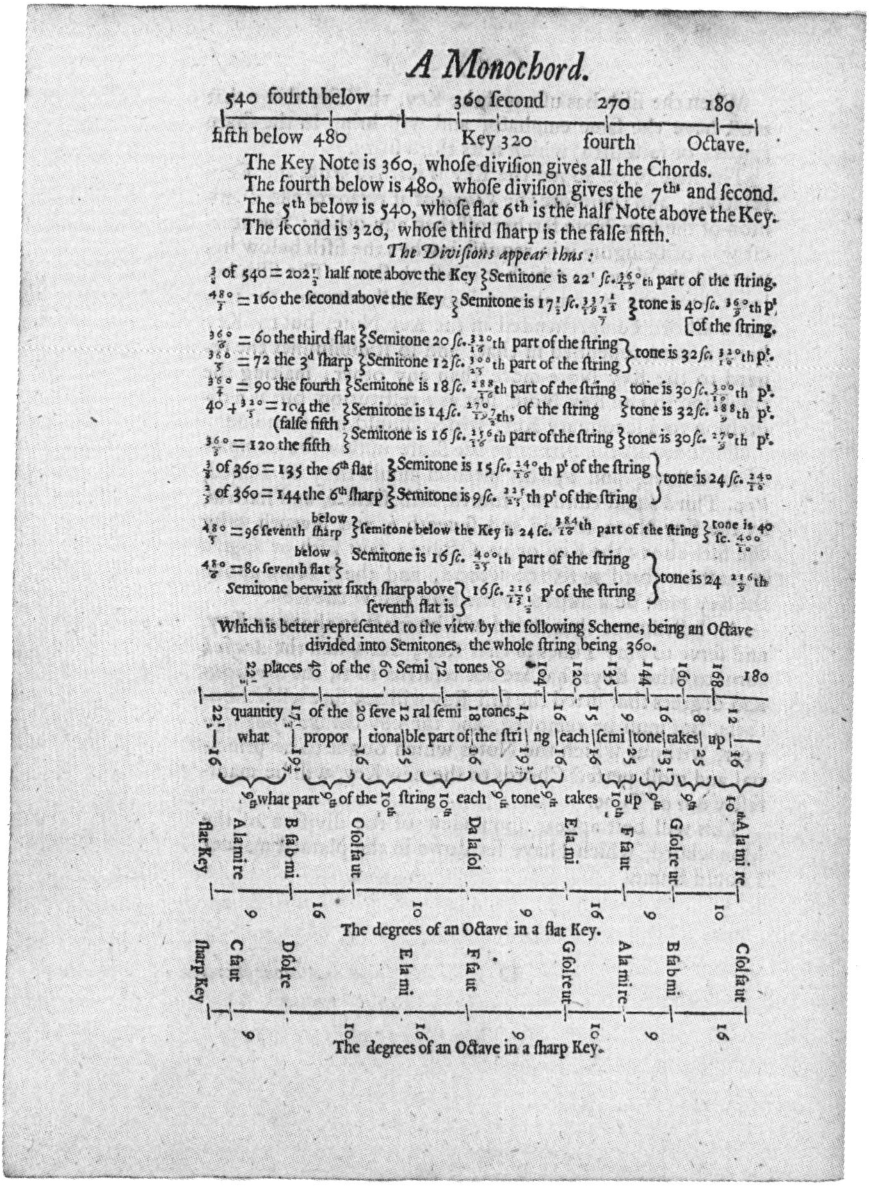

A monochord

Figure 1.C

of the tones and semitones, that lye in order above it, from what they are in relation to the true key note, which will show more gross disproportions or schismes.

Quarter notes[97] have been invented, which[,] placed in those parts where these schismes are the greatest, make the in- **[p. 28]** strument serve those keys to which otherwise it would not be in tune. Organs and harpsechords that have no quarter notes are tuned with allowances, so that there is but one perfect fifth in the compass of an octave. The rest bear on the one hand or the other, and so are indifferently well in tune to all keys, but exactly in tune to none.

Harpes, lutes, and lyra viols, that are tuned to a particular key, will be well in tune without any regard to this.

Pipes may be helped by the strength of the blast, and fretted instruments by a favourable stopping.[98]

The violin has a great advantage by not being confined to frets. But above all, the voice is most excellent, because it is not any way confined, but carried by the ear to those degrees which make exact harmony.

Of Musick Consisting of Several Parts in Consort

Musick in consort[99] consists of harmony, formality and conformity.

Harmony is the gratefull sound produced by the joyning of several tones in chord one to another.

I shewed before, that tones in chord one to another will joyn and give one entire sound, whereas discording tones will jar; and also that some chords were more perfect than other[s], according as the coincidence of their pulses was more frequent.[100] It will be proper here to say something concerning the several natures of the chords in conjunction, and the qualities of conjunct sounds.

In all conjunct sounds the base is[,] according to its name[,] the foundation, and all the upper chords are gratefull as they have relation to that, because it comprehends all the rest. And its pulses are[,] according to nature[,] bigger and **[p. 29]** stronger than those of the other parts, and so of more effect to maintain the coincidence. And for this reason, where a tune begins with conjunct sounds, the base is always the key note.

---

[97] *quarter notes*, i.e., split or slit keys added to the keyboard to improve its intonation by an increase in the number of notes in the octave.

[98] *stopping*: *pr.pple. stop* OED *v.* 27. *Mus.* to press down (a string, e.g., of a violin) with the finger in order to shorten its vibrating length and thereby produce certain intermediate sounds (c.1500–1867).

[99] *in consort*: OED *sb.*[2] *obs.* in accord, in concert (1634–1793); see also FN pp. 31.

[100] See FN pp. 7–11.

Octaves[,] as I have shewed before[,][101] do not at all alter the sound, but only swell and fill it. For the sound has its quality from the terms of the interstices between the pulses; which if they [the interstices] are larger[102] makes the sound more grave, if they be less, the sound is more acute. Now as in division,[103] the mode[104] of the time is never alter'd by doubling the number of notes in the same space of time, but the mind or memory couples them and comprehends them in the same space [of time] as if they were whole notes;[105] so is it by octaves, being [that] they increase by duplication of the pulses, in the same space of time [so that] it [duplication] produces no alteration of the sound. And though there be octaves above which are more acute, and octaves below which are more grave, so that they differ from unisons; yet as to appearance they are the same, in respect of such difference as there is between other sounds; as if a man looks upon a dye with the naked eye, and after should look upon it with a magnifying glass, though it appears bigger, yet by the figure and scituation he judges it the same.

Therefore the musick is not changed but filled by adding octaves above the upper parts. But octaves below the upper parts that should come below the base, would change the musick, by changing the foundation and basis of it, to which all the chords would have other relations than they had to the former.

A fifth is the principal chord (for the octave cannot properly be called a chord) in which there is an acquiescence; there being no other sound to which it can change to a more grateful one, and therefore is allowed in the close to fill the sound.

A fourth by its coincidence of pulses should stand in the second place, but it is not allowed in musick, according **[p. 30]** to the sweetness of its sound, because its octave below the base mends[106] the harmony so much, that it cannot be kept out of mind, but will be desired and expected. And therefore a fourth is a binding note[107] that strongly induces a close in the fifth below. So that a fourth is not for its sweetness allowed to be a chord by most musicians,

---

[101] See FN p. 10.

[102] *larger*, i.e., longer (referring to units of time).

[103] *division*, i.e., multiple notes in the time of one. For the musical art, see Simpson, *The Division-Viol.*

[104] *mode*, i.e., measure or metrical group, as duple or triple proportion; see also modes, FN p. 34.

[105] *whole notes*, i.e., a note occupying the whole of some particular time.

[106] *mends*: OED *v.* II. 7.a. *trans. rare* or *obs.* without distinct reference to defect: to improve (c.1330–1876).

[107] *binding note*, i.e., syncopated or driving note; see, e.g., Butler, *The Principles of Musick*, pp. 52, 57 (=55), 62, 64–71, 96, and Simpson, *The Division-Viol*, p. 21.

because it is not stable but subservient to a change, which change is more induced, because only a sixth can be joyned with it, which will be a third to the fifth below, and so increases the expectation of a change.

A third sharp is a chord so gratefull that it is allowed in the close to fill the sound; it being in chord to the fifth [that] bears it company. And its octave or reverse would change the musick into a sixth flat, which is the least pleasing of all chords, and therefore is not at all regarded; but the mind acquiesces in the third sharp.

A third flat, nor the sixth flat nor sharp, claim not any place in the close, because they are in discord to the third sharp and fifth, so that both cannot be allowed, and the other[s] are preferred. But they are in the body of a tune very gratefull chords that never offend the ear, nor do they invite any change by their octaves below the base, but afford a stable and pleasant harmony.

I shall not speak of discording tones which are often conjoyned in musick, because they are not allowed for their own sake, but for reasons which will appear in what I have further to say concerning consort.[108]

Wherein the second requisite is formality, which every part ought to have, though not to that degree of pleasantness that is expected from a single tune, yet in a reasonable degree.

[p. 31]

The formality of a single tune must be exact, because the composer is not confined, but may follow his fancy. But in consort the composer must consider the harmony of the parts conjunct, and therefore is not at liberty.

And there are allowances which the nature of the parts require. The formality of the treble must be airy and brisk, that of the base slow and robust. The inner parts are generally employed to fill the musick with chords, and have little curiosity[109] of formality, except upon following a point.[110] But these allowances considered, an expert musician will give a competent share of formality and sweetness to every part, making the treble move in a grave pace, and the base rest, that the beauty of the inner parts may be discerned.

The third requisite of musick in consort is conformity of all the parts. As the formality of a single part requires, that the succeeding notes should be agreeable to the former; so conformity requires that each part should have the like tendency to the succeeding notes.

---

[108] *consort*: OED *sb.*^{2} II. 3. *obs.* accord or harmony of several instruments or voices playing or singing together in tune (1587–1695); see also FN p. 32.

[109] *curiosity*: OED *sb.* II. 9. *obs.* elaborateness, elegance, artistic character (c.1380–1697).

[110] *point*, i.e., a motif, theme or subject suitable for treatment by the techniques of variation or, more commonly, imitation; see also points, FN p. 33, and Simpson, *The Division-Viol*, especially pp. 53–56.

Which tendency is many times so regarded, that the present harmony must yield to it. For if the succeeding notes are of great relation to the key, and the formal way to them in the several parts is by discording notes; nevertheless that way which gives so good expectation must not be balked, but rather the discords will be allowed, whose harshness is taken off by the expectation of the succeeding chords, and the harmony of the following notes is the more welcome for the badness of the passage.

Discords are allowed in musick upon this score, and they are also in other respects of excellent use; as when the mind is to be carried from any particular note, it may be affronted by a discord; or if a note in small esteem be to **[p. 32]** be favoured, some other note may come in that agrees with that, and is discord to the rest, and so may draw the mind to the regard of that note which it strengthens and supports. Infinite other instances may be made, but my business being only to illustrate the reasons I go upon, and not to teach rules, these shall suffice.

From what I have said, it appears, that when each part has its kind and proportion of formality, and the parts are in chord to the base, and not in discord to one another, but carry all the chords to fill the harmony, when they have a like tendency to their respective succeeding notes, when they imitate one another's measure, then is the consort compleat. For the ear is pleased with the harmony, and the mind is amused and entertained to observe the particular parts[,] how they dance to and from the key, and from one key to another, how they hunt one another, and in a manner imitate humane passions.

To the end the hearer may the more indifferently observe the contexture of the parts, it is necessary that they be fairly interwoven, one part must not cling to another, by moving together in eighths or fifths which would draw the attention too much. For the same reason they may not jump into eighths, which I have heard called, hitting an eighth in the face.[111] But an equall hand should be held over all the parts, and the sweetness distributed amongst them.

I will not go about to describe excellent musick, which would require a poet as well as a musician. I am only in the part of a philosopher, to show what is allowable musick, in order to make the reasons[,] upon which their rules are founded[,] understood.

And that I have done, as it were by a glimpse, knowing the difficulty of being exact upon such a subject. He that **[p. 33]** desires to be more fully informed concerning it, may have recourse to other large treatises that have been written of musick by authors of our own, as well as other nations.

---

[111] See Morley, *A Plaine and Easie Introduction to Musick*, p. 75: '*hitting the eight on the face*', glossed by Simpson, *A Compendium*, pp. 96–97, thus: 'that is, when an upper Part meeting the *Bass* upon an 8*th.* doth skip up from thence into some other Perfect Concord'. See also RN, *The Musicall Grammarian*, p. 163, who repeats FN.

There are tricks[112] in musick, as fugues, points, descant upon plainsong, cannons, etc., the former, namely, fugues and points are extream delightfull, and embellish the musick, carrying the attendance of the mind from one part to another, and restrain not the fancy of the composer, but advance it.

The other tricks that are more elaborate, are to teach or shew mastery in composing, rather than to please.[113] For such fetter'd musick is alwayes worse, and the hearer shall not observe the art till he be told of it. It is like acrostick verse, or rhithm[114] in plays, which shew skill or labour, but serve not those ends for which the pieces were designed.

OF TIME, OR THE MEASURES OF MUSICK

As the musick and harmony of tones depends upon the regular distance of the pulses, though the interstices are not perceived, nor can be measured by us, so when the interstices are so large that they can be measured by our sense, the observance of a regular proportion of time affords great pleasure; as in the beating a drum, or striking a pair of castanets.

The due observance of time is gratefull for the same reason that I gave for the formality of a single tune,[115] because the subsequent strokes are measured by the me- **[p. 34]** mory of the former, and if they do comprehend them, or are comprehended by them, it is alike pleasant. The mind cannot chuse but compare one with the other, and observe when the strokes are coincident with the memory of the former.

Therefore the less the intervals are, the more grateful is the measure, because it is easily and exactly represented by the memory, whereas a long space of time that cannot be comprehended in one thought, is not retained in the memory in its exact measure, nor can abide the comparison, the time past being alwayes shortned by so much as it is removed from the time present.

The modes of time are very few, being only in duple or triple proportion, and the diminishing is alwayes in geometrical proportion,[116] as two, four, eight,

---

[112] *tricks*: OED *sb*. 6.a. *concr. obs. rare* clever contrivances or inventions (1548–1601).

[113] Perhaps by 'other tricks' FN means the more erudite techniques of canonic writing used, e.g., by Morley, *A Plaine and Easie Introduction to Musick*, not only as a pedagogic device but also as an exemplar of the art of a 'great master'.

[114] *rhithm* = *rhythm*: OED *sb*. I. 1.a. *obs*. riming or rimed verse (c.1557–1695); cf. *rhyme*: OED *sb*. 2. verse marked by consonance of the terminal sounds (1652–1885).

[115] See FN p. 23.

[116] *geometrical proportion*, i.e., geometrical progression, a series in which the ratios between successive quantities is constant.

sixteen; that it might serve not only to the basis of the time, but also to the other intermediate degrees of diminution, which are in memory, if not in play.

The triple can give two spaces to one, the duple can give three spaces to one, and although the duple or plain time seems to have spaces of even length, yet there is loudness and briskness given to every other pulse, which makes it eminent, and alwayes happens upon coincident strokes. This difference is between the fore-stroke and back-stroke in musick,[117] where the notes are marked to be of an even length.

These disparities of even and odd, and of two to one, and three to one are necessary, not only where time alone is regarded, as in a drum, or upon a horn, to show the coincident strokes; but [also] in respect of musick. For thereby length and loudness is given to the key notes, and **[p. 35**[118]] such other notes as should bear an emphasis; and the other notes that have the short or odd foot, seem but steps and passages to them, and indeed are so little regarded, that whether chord or discord they must be taken if they lye in the way. In the same manner, in all shaking graces[119] [where] the hovering is unequal betwixt two notes, the advantage is given to the true note, and the other has but faint touches, to make preparation for that which is to succeed.

[Conclusion]

It is now time to put bounds to this discourse, which by reason of the method I engaged my self in, is become much longer than I intended. I will only add these two observations:

1. It appears plainly how musick comes to be so copious. For considering the species of keys, the number of them, the variety of chords, the allowable mixing of discords, the diversity of measure, it is not to be wondred at, that it should[,] like language[,] afford to every age, every nation, nay, every person, particular stiles and modes.[120]

---

[117] *fore-stroke and back-stroke*: OED *obs.* a forward [and backward] stroke in bell-ringing and in sword play (1674–1779). Here, the motion of the hand in beating time, which Simpson, *A Compendium*, pp. 14–15, described as

> *Down* and *Up*, successively and equally divided. Every *Down* and *Up* being called a *Time* or *Measure*. And by this we measure the length of a *Semibreve*; which is therefore called the *Measure-Note*, or *Time-Note*.

[118] 1677 text has '25'.

[119] *all shaking graces*, i.e., varieties of the trill, a more or less rapid alternation of the main note with the one a whole or half step above it; for the varieties, see Simpson, *The Division-Viol*, pp. 11–12.

[120] *modes*, i.e., fashions.

2. It appears that tones[121] or modes of musick in antient time could not be of other kinds than are now; for there can be no other in nature. Therefore the great effects it then had, if truly related, must be imputed to the rarity of it, and the barbarity of the people, who are not transported with any thing after it becomes common to them.

Finis.

[121] *tones* = *tonus*, i.e., the Latin term for the Greek modes.

A Monochord

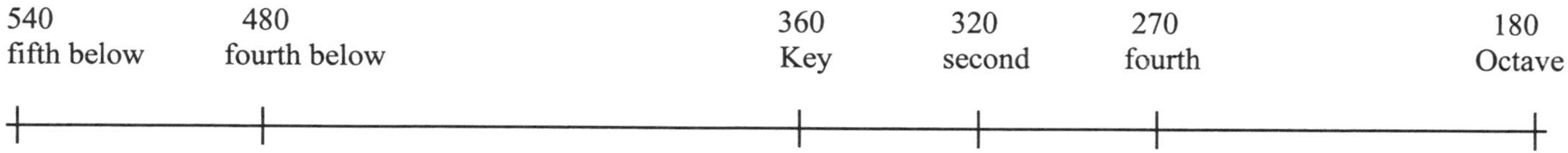

The key note is 360, whose division gives all the chords.
The fourth below is 480, whose division gives the 7ths and second.
The 5th below is 540, whose flat 6th is the half note above the key.
The second is 320, whose third sharp is the false fifth.

Clarification of *Figure 1.C*

Figure 1.D

Figure 1.D, continued

The divisions appear thus:

⅜ of 540	= 202½ half note above the key	Semitone is 22½	*sc.* 360/16th part of the string.	Tone is 40 *sc.* 360/9th part of the string.
480/3	= 160 the second above the key	Semitone is 17½	*sc.* (337½)/(19 2/7)th pt string.	
360/6	= 60 the third flat	Semitone is 20	*sc.* 320/16th part of the string.	Tone is 32 *sc.* 320/10th part of the string.
360/5	= 72 the 3d sharp	Semitone is 12	*sc.* 300/25th part of the string.	Tone is 30 *sc.* 300/10th part of the string.
360/4	= 90 the fourth	Semitone is 18	*sc.* 288/16th part of the string.	Tone is 32 *sc.* 288/9th part of the string.
40 + 320/5	= 104 the false fifth	Semitone is 14	*sc.* (270)/(19 2/7)th pt of string.	Tone is 30 *sc.* 270/9th part of the string.
360/3	= 120 the fifth	Semitone is 16	*sc.* 256/16th part of the string.	
⅜ of 360	= 135 the 6th flat	Semitone is 15	*sc.* 240/16th part of the string.	Tone is 24 *sc.* 240/10th part of the string.
2/5 of 360	= 144 the 6th sharp	Semitone is 9	*sc.* 225/25th part of the string.	
480/5	= 96 seventh sharp below [the key]	Semitone† is 24	*sc.* 384/16th part of the string.	Tone is 40 *sc.* 400/10th part of the string.
480/6	= 80 seventh flat below [the key]	Semitone is 16	*sc.* 400/25th part of the string.	Tone is 24 *sc.* 216/9th part of the string.
Semitone betwixt sixth sharp above [and] seventh flat is 16			*sc.* 216/(13½)th pt of string.	

† below the key

Which is better represented to the view by the following scheme, being an octave divided into semitones, the whole string being 360.

places of the semitones [↓]

[0] 22½ 40 60 72 90 104 120 135 144 160 168 180

22½ 17½ 20 12 18 14 16 15 9 16 8 12

quantity of the several semitones [↑]

what proportionable part of the string each semitone takes up [↓]

16 19 2/7 16 25 16 19 2/7 16 16 25 13½ 25 16

⇓ 9th ⇓ 9th ⇓ 10th ⇓ 10th ⇓ 9th ⇓ 9th ⇓ 10th ⇓ 9th ⇓ 9th ⇓ 10th

what part of the string each tone takes up [↑]

Figure 1.D, continued

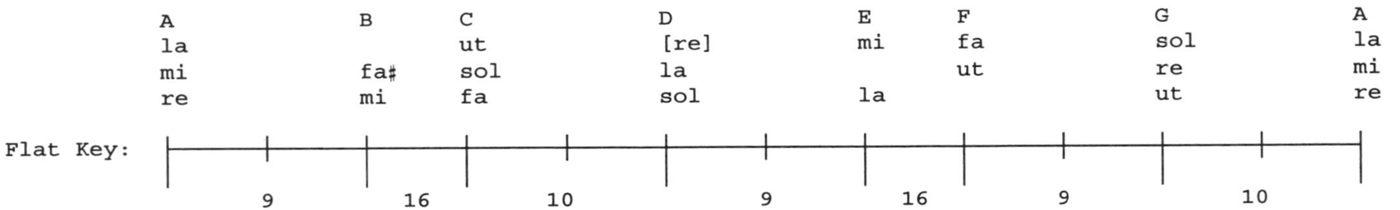

The degrees of an octave in a flat key.

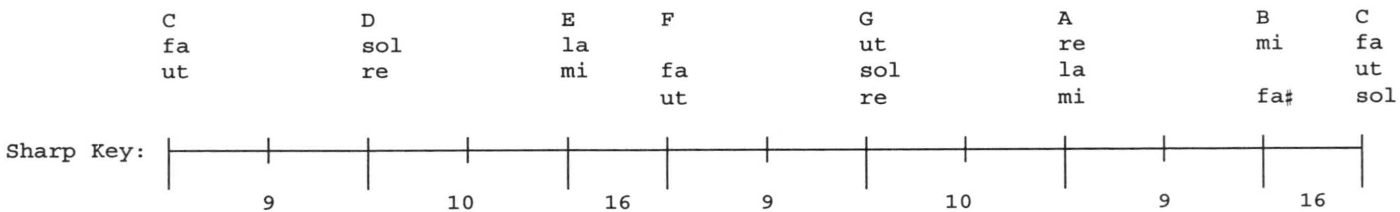

The degrees of an octave in a sharp key.

2

ANONYMOUS NOTICE

PHILOSOPHICAL TRANSACTIONS OF THE ROYAL SOCIETY
Vol. 12, No. 133 (March 25, 1677), pp. 835–838

[p. 835]
IV. *A Philosophical Essay of Musick*: London, printed for J. Martyn, Printer to the R. Society, at the Bell in St. Paul's Church yard, 1677, in 4°.

This author's design being to explain the nature of musick, he begins to inquire into the cause of sound. In order whereunto, he considers some of the chief phænomena of sound, as 1. that it may be produced, according to him, in the Torricellian vacuity; 2. that it causes motion in solid bodies, and is diminished by the interposition of solid bodies; 3. that if the bodies interposed are very thick, its passage is wholly obstructed; 4. that it seems to come to the ear in strait lines when the object is so scituated that it cannot come in a strait line to the ear; 5. that when the air is not in motion, its extent is spherical; and when there is a wind, the sphere is enlarged on that part, to which the wind blows, and diminished on the contrary part; 6. that it arrives not to the ear in an instant, but considerably slower than [light to the organ of] sight; 7. that it comes as quick against the wind as with it, though not so loud nor so far.

Hence he raises the following hypothesis. He supposes the air, we breath[e] in, to be a mixture of different minute bodies which are of different sorts and sizes, though all of them are so small as to escape our senses. The grosser of them he makes elastical, and such as are resisted by solid bodies, altogether impervious to them. The smaller parts pass through solid bodies, tho not with that ease but that upon a sudden and violent start of them, they shock the parts of solid bodies that stand in their way, and also the grosser parts of the air. Lastly, [he conjectures] that there may be another degree of most subtile ethereal parts, with which the interstices of these and all other bodies are repleat, which find freer passage every where, and are capable of no compression, and consequently are the medium and cause of the immediate communication of [light to the organ of] sight.

Now, of these three, he esteems the middle sort to be the medium and cause of sound, and that at any time, when the grosser air is driven off any space, and leaves it to be possest by these **[p. 836]** and other more subtile bodies, and returns by its elasticity to its former place, then, are these parts extruded with violence as from the center of that place, and communicate their motion as far as the sound is heard. Or, when any solid body is moved with a sudden and violent motion, these parts must be affected thereby. For, as these parts are so much resisted by solid bodies as to shock them; so, on the contrary, they must needs be moved by the sudden starting of solid bodies.

So that (according to him) sound may be caused by the tremble of solid bodies without the presence of gross air; and also by the restitution of gross air, when it hath been divided with any violence. Thus, (saith he) we see, that a bell will sound in the Torricellian space. And, when the air is divided with any sudden force[—]as by the end of a whip having all the motion of the whip contracted in it, and by a sudden turn throwing off the air; or by accension, as in thunder and guns; or by any impression of force carrying it where other air cannot so forcibly follow, as upon compressing of air in a bladder till it breaks, or in a pot-gun [—]a sudden crack will be caused.

Having laid down this hypothesis, and left his reader to apply it to the afore-mentioned phænomena, he proceeds to the discourse of musick it self, and maketh it a considerable part of his business to shew, how this action that causes sound, is performed by the several instruments of musick; having taught his reader, first, what a tone is, and that the tones useful in musick are those within the scale, in which they are placed as they have relation to one another. Secondly, wherein consists that relation of tones and the union of mixt sounds. Which done, he explains, how tones are produced, and what assistances are given to the sound by instruments. Where he teaches, that wherever a body stands upon a spring that vibrates in equal terms, such a body, put into motion, will produce a tone, which will be more grave or acute, according to the velocity of the returns. Wherefore strings vibrating have a tone according to the bigness or tension of them; and bells that vibrate by cross ovals, produce notes according to the bigness of them, or the thickness of their sides; and so do all other bodies, whose superficies, being displaced by force, results or comes back by a spring which carries it beyond its first station. And here, to make it to be understood, how every pulse upon such vibrations causes sound, our author gives us to consider, that the gross air is thrown off by the violence of the motion, which continues some moments of **[p. 837]** time after the return of the vibrating body; whereupon some space must be left to the subtil matter, which upon the resilition of the air starts as from a center; which action being the same, by the author supposed to be the cause of the sound, is repeated upon every vibration.

But finding it more difficult to shew, how tones are made by a pipe, where there are no visible vibrations; he considers the frame of a pipe, and the motion of the air in it, and thereby attempts to find the cause of the tone of a pipe, and the pulse that gives the sound[,] not omitting to explicate, how tones are made in violins, harpsecords and dulcimers.

To this he subjoyns an ingenious discourse of the varying and breaking of tones, endeavouring to explain, how it is caused both in strings and pipes;

where occur divers pertinent observations concerning the motion of pendulums, the trumpet marine, and the true trumpet, as also the sackbut. And having shew'd, that sound doth cause a motion not only of solid bodies, but of the grosser parts of air within the sphere of it; he considers, that if the air, which is moved by being inclosed, stands upon such a degree of resistance to compression, that it hath a spring vibrating in the same measure with the sound that puts it into motion, there will be the same [resonance] effect, as when 2 strings are tuned in unison; that is, the motion will be so augmented by succeeding regular pulses, that the inclosed air may be brought to ring, and produce a tone. Where he taketh notice of the advice of Vitruvius in his *Architecture*, importing, that in the structure of a theatre, there should be vases or hollow pots of several sizes to answer all the notes of musick placed upon the stage in such manner, that the voice of them which sang upon the stage might be augmented by the ringing of them; Vitruvius mentioning divers antient theaters, where such were, in some of brass, in some of earth [i.e., clay].

After this, he descends to the consideration of the nature of keys in musick, and of a single tune; which latter, he saith, consists in the succeeding notes having a due relation to the preceding, and carrying their proper emphasis by length, loudness and repetition, with variety that may be agreable to the hearer.

Next, he treats of schismes and the scale of musick; shewing that this scale is not set out by any determinate quantities of whole or half notes, though the degrees are commonly so called; but that the degrees in the musical scale are fixed by the ear in these places, where the pulses of the tones are coincident, with- **[p. 838]** out any regard to the quantity. Here he endeavors to shew, how all the notes come into the scale by their relation and dignities; whence he thinks it will be obvious, why, for easiness of instruction and convenience, the scale of degrees of musick is made as musicians now exhibit it.

Having dispatched that work, he proceeds to musick that consists of several parts in consort, which is made up of harmony, formality and conformity. Of which, harmony is the grateful sound produced by the joyning of several tones in chord to one another. Formality requires, that the succeeding notes be agreable to the former; and conformity will [require], that each part have the like tendency to the succeeding notes.

Lastly, he speaks of time or the measures of musick; the due observance of which is grateful for the same reason given for the formality of a single tune, because the subsequent strokes are measured by the memory of the former, and if they do comprehend them, or are comprehended by them, it is alike pleasant; the mind cannot chuse but compare one with the other, and observe when the strokes are coincident with the memory of the former. Whence it is, that, the less the intervals are, the more grateful is the measure, because it is easily and exactly represented by the memory; whereas a long space of time that cannot be comprehended in one thought, is not retain'd in the memory in its exact measure, nor can abide the comparison, the time past being always shortned by so much as it is removed from the time present.

The whole is concluded by two observations, by which we shall likewise conclude this account: 1. That it plainly appears by the discourse of this tract, how musick comes to be so copious; for, considering the species of keys, the number of them, the variety of chords [i.e., accords], the allowable mixing of discords, the diversity of measure; it is not to be wondred at, that it should, like language, afford to every age, every nation, nay, every person, particular stiles and modes. 2. That it appears likewise, that tones or modes of musick in ancient time could not be of other kinds than are now, since there can be no other in nature. Wherefore the great effects it then had, if truly related, must be imputed to the rarity of it, and the barbarity of the people, who are not transported with any thing after it become common to them.

Imprimatur,
BROUNCKER, P.R.S.

March 29. 1677

3

ISAAC NEWTON

LETTER TO JOHN NORTH

[f. 1]

Cambridge April 21. 1677

Sir

The esteem you express of my judgment I must impute to your goodness who are willing to make the best of every thing. Yet since it is your desire to have my opinion about the new treatise of musick, I shall give it you, though perhaps not so largely as you may expect, there being some things which I cannot speak positively to for want of experiments and skill in musick.

Page 5 line 14[1] and page 6 line 30.[2] The author asserts that sound is produced in the Torricellian vacuum, and thence seems to collect[3] that the medium of sound is not the grosser air but some subtler aerial fluid of a middle nature between that air and æther, which can penetrate glass and other gross bodies. But it is to be suspected that this experiment of the Torricellian vacuum holds only when the glas is not well emptied of air. For Mr Boyle (Experiment 27)[4] repeating it with a watch hung in his receiver,[5] found that as the receiver was more and more emptied, the noise made by the ballance wheel grew fainter and fainter till at last it was not heard at all though the handle and wheels of the watch were still seen to continue their motion as freely as at first. Yet the louder sound of a bell continued audible when the air was drawn out, though perhaps it would not have done so could the receiver have been fully exhausted of air, and the bell have been susteined in the receiver by something which might not touch the glas[.] To the best of my remembrance I have also some

---

1 FN p. 5: '1. It [sound] may be produced in the Torricellian vacuity.'

2 FN p. 6: 'We see a bell will sound in the Torricellian space.'

3 *collect*: OED *vb.* 3.c. *rare* to recall to remembrance, recollect (1610).

4 Robert Boyle, 'Experiment XXVII', *New Experiments physico-mechanicall, touching the Spring of the Air, and its Effects*, (*made for the most part, in a new pneumatical Engine*)...(Oxford, 1660; 2d edn., with additions, 1662). For details, see Chapter 2, pp. 42–44 *supra*.

5 *receiver*: OED *sb.* 5.a. *Chem.* b. bell glass of an air-pump (1660–1816).

where read of an alarm watch whose alarm being made to go, the sound of the bell grew faint by drawing out the air.[6]

Page 5 line 18.[7] The author asserts that a sound seems to come in streight lines to the ear though an obstacle be situated between the ear and the sounding body so that it cannot come in **[f. 1v]** streight lines. But this I doubt of[,] if[8] he mean that it seems to come in streight lines from the sounding body. If indeed the interposed obstacle be not too gross and compact, suppose a glass window or thin wall of wood or mortar, the air may by shaking it propagate the sound through it, and then the sound will be heard in a streight line from the sounding body; but if the obstacle be massy, suppose a very steep and high hill or a sollid high wall of brick or stone, I am apt to think the sound will seem to come from the top of the hill or wall rather then in a direct line from the sounding body behind it. And some such diverting of sounds I have observed occasionally, in walking on a street close by a single house whilst bells were ringing on the other side. The house to the best of my remembrance, had no windows on that side next me and the sound of the bells seemed to come from that end of the house which I was nearest to though the bells were directly behind it. So in a room of stone walls with but one window, the sound will seem to come from the window though the sounding body without lye not that way.[9]

Page 8 line 14.[10] The author affirms that, where the vibrations of two sounding strings are equal, they will work one another to a coincidence or synchronism, and on this position grounds all his discours of concords and discords, affirming that unisons alway[s] strike the ear together, octaves at every other puls of the treble, fift[h]s at every 3^{d} pulse, etc. But it seems otherwise to me. For though the pulses of the strings should reduce one another to a synchronism, yet those of one string will strike the ear sooner or later then those of the other, accordingly as the ear is more or less near to one string then to the other.

Since sounds are not propagated in a moment, the pulses must not be supposed to extend each of them at once from the sounding body to the ear in such manner that all the air in that interval be moved together first forward and

---

6 Otto von Guericke, *Experimenta nova (ut vocantur) Magdeburgica de vacuo spatio* (Amstelodami, 1672), Book 3, Chapter 15, pp. 91–92.

7 FN p. 5: '5. It [sound] seems to come to the ear in strait lines when the object is so scituated that it cannot come in a straight line to the ear.'

8 MS has 'if,'.

9 This sentence, inserted below the previous sentence, is completed, sideways, in the margin of f. 2. Hence, it seems to have been included as an after-thought.

10 FN p. 8: 'Upon this reason [resonance], wheresoever [the terms of] the vibrations of two sounding strings are equal, the pulses will be coincident.'

then backward, and so forward and backward again so long as the sound lasts. But the pulses are rather to be conceived like so many spherical concentrick waves whose center is the sounding body and which arising continually from that center dilate and flow on from thence with that swiftness we find sound propagated till they arrive at the ear; new pulses from the center continually succeeding them after the manner of undulations made by throwing a stone into water. Let therefore A and B [in *Figure 3.A*] represent two

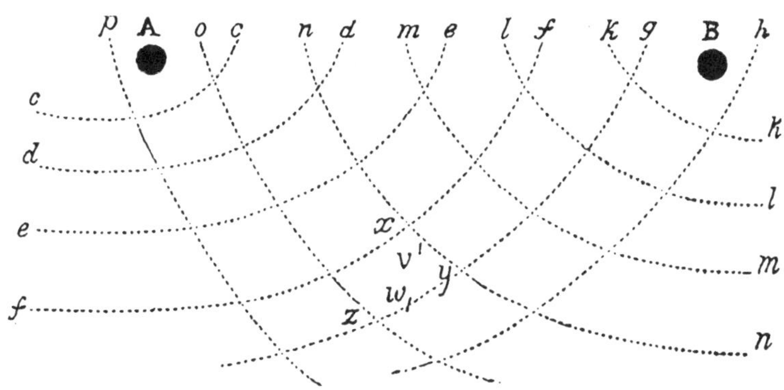

Representation of sound reception from two tonal stimuli

Figure 3.A

sounding bodies of like tones, c, d, e, f, etc., the pulses continually propagated from A; and k, l, m, n, etc., the pulses propagated from B. And if the ear be placed at x, or y, or z, or any place where the pulses at any line intersect, they will strike the ear at the same time, but if it be placed between x and y suppose at v, the puls xn which comes from the body B will strike it in the intervall of the pulses xf and yg which come[11] from the body A. And the like will happen if the ear be placed between y and z as at w, or in any **[f. 2]** pulse of one body and between two pulses of the other. So that the ears of two men or the two ears of the same man according to their position from the sounding bodies may be struck[,] the one at once the other successively by the pulses, and yet in all positions of the ear the sounds and their harmony are heard the same. The like

[11] Previous transcriptions of the letter have 'come', but the MS here is indecipherable, since the word originally written ('raine'?) has been written over.

of bodies tuned to a fift[h] or any other concord. Seing therefore the pulses of two sounds may at the same time strike one ear both together and another ear alternately and yet do exhibit the same concord to all ears, it follows 1st that concords arise not from the coincidence of pulses at the ear nor have any dependance on such coincidences, and 2dly that unisons are rather a harmony of two like tones then a single tone made more loud and full by the addition as the author would have it, page 8, line 20.[12]

The explication of the sound of whistles page 12 is very ingenious,[13] but I fear not altogether substantial. Yet for want of apposite experiments I can neither sufficiently confute it nor confirm any new hypothesis.

The discours also about breaking of tones into higher notes seems very ingenious and judicious, but I want experience to discern whether altogether solid, and much more do I want experience and skill to enable me sufficiently to judge of what follows about tunes, the scale of music, and consort;[14] this requiring a combination of musical and mathematical skill, and therefore I shal content my self with having thus far animadverted upon the author.

I am much obliged to you for giving me notice of the objection made against my notion about colours. But the experiment succeeds otherwise then tis reported. If you place your eye where the blew light falls on the wall so that a by-stander see your eye of a blew colour you will at the same time see the prism of a blew colour; and if you place you[r] eye in the red light you will see the prism red. What colour a by-stander sees fall on your eye you will see at the prism[,] as I can affirm by iterated experience.

The last week I called at your lodgings and hope ere long to have an opportunity to wait on you again. In the meane time I rest with my thanks to you for your kind acceptance of my former letters.[15]

Your humble servant and honourer
Is: Newton.

---

[12] I.e., 1677 text, lines 19–23; see FN p. 8: 'Where the terms are equal [i.e., 1 to 1], and the coincidence is at every pulse, it may be said [to be] the same sound though proceeding from several instruments. For the mixed sound has no alteration, but by being more loud and full by the addition.'

[13] At p. 12, FN treats sound generation in pipes, and at p. 14, in whistles (a species of flue pipe).

[14] See FN pp. 21–33. For Newton's student notes on some of these topics, see (GB:Cu) Add MSS 3958 (B): f. 31; 3970(1): ff. 12–15; 4000: ff. 104–113, 137v–143, and 3970(1): ff. 12–15. For a description of some of the contents, see Gouk, *Music, Science and Natural Magic*, pp. 230–237.

[15] No other letters to JN are printed in Newton, *The Correspondence*, perhaps for the reason RN gave in *Life/FN*, p. 98, that 'the doctor [JN] made him [his executor, FN] promis to burne all books, and papers he found of his owne hand wrighting; as he [FN] most punctually performed'.

4

Roger North

Some Notes upon an Essay of Musick printed, 1677, by way of Comment, and Amendment

British Library Additional Manuscript 32531: ff. 42–52

[f. 42]

Of sound[1]

Page 1.[2]

Certain phænomena of sound are enumerated, viz., 7. that in a wind, the sphear of sound inlargeth to leeward and shrinketh to the windward, and 9. that the sound moves as quick to windward, as the other way but not so loud nor so farr.

The manner of this action is easily conceivable if the circles made in water, which runneth in a stream[,] be considered; for there is no difference in the phenomenas of the one and the other. And as I take it the circles will dilate truly sphericall, but *inter dilatandum*,[3] the water being carry'd off by the stream, the circles, as well as their center shall be removed downe the stream. So he that stands upon the bank and observes, shall preserve the idea of the puls, as it fell with respect to him. Now the place of the sound created, which is like the puls, or center of water circules, removes in a wind to leeward, and when the sound is heard a furlong to the windward, it may be the center is removed to the leeward, a quarter furlong; and then a furlong to leaward of the stroke hath it, a quarter neerer the center of the circle; which makes it seem louder. But in truth in all parts equidistant from the center of the dilatation the sound is equally strong, which is what I suppose the author meant, but his expressions are somewhat too concise.

Page 6.

The author speaking of sound supposeth it to consist in an elastick air wee breath[e,] which doubdtless is true, of which I shall speak in the close of

---

[1] MS has '(Of sound)' placed at the left.

[2] *Page 1* = FN p. 5.

[3] *inter dilatandum*: (Latin) during the dilation.

this chapter.[4] But he restraines it to one sort, viz., what he calls the midle sort. Now I take all the air wee breath[e], to be elastick, or springy.[5] And even the ether (as wee style it,[)] which **[f. 42v]** fills the derelicted space in the Torricellean experiment, is elasticall. For the mechanicall, and as I thinck the true notion of elasticity is onely, an impediment from compression to a motion that would be freer if the impediment were removed. And that will be in aeriall matter, till you come to parts perfectly sphericall, if not the most minute. For if they are any wais angular they will strike each other in turning which produceth the effect of expansion, that wee call elastick or springy.

Now it is possible, and I beleev true, that to some elastick matter, glass is pervious, and to other, not. And in the barometer, what cannot get thro the glass, which, as the author speaks, is the grosser part, is excluded. But if a body could be found less pervious then glas, the barometer within the barometer might be erected, as [in *Figure 4.A*] B within C and A within B and so *ad infinitum*. But our observation is confined to the grosser, that is[,] wood, glass, and the like, but wee are not thence to conclude, that ether is not elastick, while it is under the same compressure as other air is.

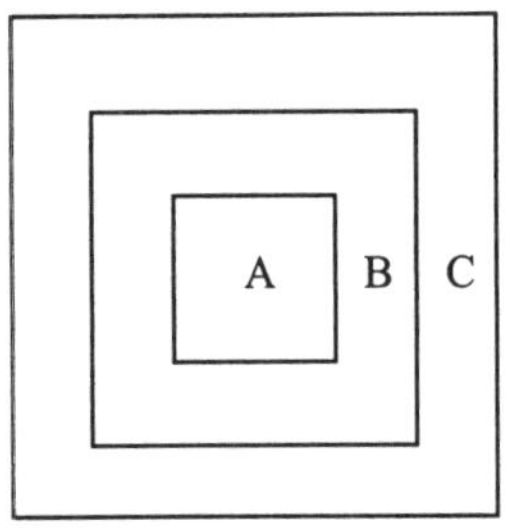

Barometer within a barometer

Figure 4.A

And this facilitates the solution of sound, as the author means, for els it would not be in a baro[s]cope-vacuity.

The author supposeth that upon every sound the air is devided, and quasi a Torricellean space made, and that the close of the air, is that which makes the crack or sound.

---

[4] See RN ff. 43–43v.

[5] RN uses various forms of the word 'spring' to mean: *(a)* pressure or expansive force (of the air), ff. 42, 42v, 43v, 48v; *(b)* a spiral wire, ff. 43v and (*fig.* of the air) 51; *(c)* tension or stretching force (of a musical string), ff. 49, 49v; and *(d)* originating source, f. 46.

I thinck this doth not clearly enough explain the caus of sound, and it may be goes farther then is needfull for the fetching of it. Therefore take my opinion of it.

Our senses are facultys limited with respec[t] to circumstances. For wee that can see minute nebulæ[6] of starrs, and hear many faint eccos of sounds in a still night, when the sun, moon and greater noises, that the day is filled with are absen[t,] **[f. 43]** cannot discerne or hear either [stars or echoes] at noonday. So that wee are to conceiv that in sound, considering how many multitudes of different sounds wee perpetually live in, [it] must come up to a certein degree of strength and force to be perceived from the rest, which, if slight it will be blended and drowned.

Now in the matter of sounds, I will not define how weak a movement in the air may give us such a sensation as that is, becaus I have no measure of the capacity of sence. But I must suppose, that unless a movement be made with a vivacity to a certein degree, such as in most ordinary sounds wee are acquainted with, wee shall not perceiv, or distinguish it from others so as to call it sound. And it must be granted that as the silence is, so fainter motions are perceived in sound, which every one knows that is used to observe in the night, when most things are at rest.

This premised, I conceiv that whenever a motion is given to the air, in any place, which occasions a compression of it, or works upon its elasticity[,] there is that which gives us the sence of sound, *if it* [the motion] *be strong, or quick enough*. Every movement of the air is not sound, as a gentle brees, or wind, nor if it be never so violent; becaus it goes all together, and, at the foramen of the ear meets with a stop.[7] Besides if it were so forcible to carry its activity to the timpanum, the world would be ruined by the hurricane of it. But when a compression is made, that[,] from the elasticity of the body[,] will compress the next, and so the next, expanding in circle, weakning as it goes, and turning about corners, where is not a force[d] passage; and passeth in to the foramen of the ear thro the ductus[8] of air there, and makes a sence [impression] upon the timpanum, from the quickness of it.

[f.43v]

So that I doe not conceiv there is any need of deviding the grosser [part] of the medium, or that the return of it, should be needfull to make sound if the quickness of it be otherwise sufficient, as in some of the author's instances it is,

---

6 *nebulæ* (pl. of nebula), i.e., an indistinct cloud-like cluster of distant stars; see also RN ff. 44, 45. OED 3. *Astron.* gives the first use as 1727.

7 *stop*: OED *sb.*[1] II. 7. *obs.* an impediment (c.1508–1725); see also RN f. 43v n.10.

8 *ductus* = *duct*: OED *sb.* 6.a. *Phys.* a tube or canal in the animal body through which bodily fluids are conveyed (1667–1966). Here, the air itself as a tube, i.e., column.

as that of thunder, bladder, potgun,[9] etc.[10] But whenever the air is devided by accensions,[11] as in gunns and (it may be) thunder[,] the force of the spring is so great to returne, that the meeting may be, and is such a choke to[12] the air, as sends out a violent expanding sound [wave].

Divers experiments may be had or fancied to illustrate this notion of sound; as if a wier-worm, such as [may be] use[d] to represent a snak[e] in a box[,] were made of a great length, and a force came upon one end, and compres't the spring, that compressure should run thro the whole length, and returne backwards and forewards, till it dys. So where cords, or ropes of great weight are strained, which wee may find where buildings are, if a stroke be given at one end, it will sensibly pass to the other, and so back, returning divers times, till it insensibly dys; and as the rope is smaller and lighter, so is the motion quicker. And the returnes may give a representation of an ecco but I rather thinck that an ecco, is a new sound derived out of the former by an aceurs[13] or contraction of the movement; as light contracted burnes, and hath a greater force then the generality of it.[14] As in vallys, ruines, and cavitys, the movement may be contracted, and acquire a new force. And when divers of these, are opposed, the ecco shall repeat, like reflecting glasses as to light, till the image dys insensible.

So that, to conclude my observations upon this chapter, I am of opinion that, sound is no other, then a stroke upon the air, occasioned by the movement of some solid body in it, which produceth a compression and gives a quickness

---

[9] *potgun* = *popgun*; see FN p. 6 and n.16.

[10] In the left-hand margin:

Memorandum. Canon, is sensible to our bodys as an heav, so also breaks windows which is a perpetuall thrusting of the air forewards, from a mighty compression at the gun, and I beleev is from the returne of the air into the peice, which meeting with a stopp, is compressed within the metall, and reflects out in circles.

[11] *accensions*: OED *sb. obs.* or *arch.* the actions of kindling or of being kindled; ignitions (1646–1801).

[12] *choke to*, i.e., shock to. Here, elastic impact (compression); see FN p. 6 and n.13.

[13] *aceurs* = *occurse*: OED *obs.* meeting (1621–92). Here, the point where the vibrating motion of the body (spiral spring or cord) meets, and is reflected back from, the rigid support. For RN's other uses of 'occurs', see f. 72 of 'The Theory of Sounds...', (GB:Lbl) Add MS 32534: ff. 1–82v, and f. 92v of 'Theory of Sounds...1728', (GB:Lbl) Add MS 32535: ff. 74–149.

[14] In the left-hand margin:

Sound differs from wind, becaus in that the whole air moves, without compression. But the hollows and corners that it meets with and vorticates in, produce compression, which makes a whistling.

[Sound differs] from the action of light, which, wherever it is, is instantaneous, and affects the whole medium at once, which sound doth not.

vorticates: (Northism) *v.* creates a vortex.

of motion proportione[d] to our sence [of hearing], in presenting that wee call sound.

[f. 44]

OF TONE

It is the vnhappyness of musick, that the vse and practice of it, seldome meets with learning and ingenuity in one person. Wherefore it hath bin ever a mistery, and litle understood, as to the principles and reasons of it; and when ingenious men and scollars have apply'd themselves to consider it, after all their paines which hath bin very great, wittness Mersennus, Wallis, Vossius,[15] the result hath visibly wanted acquaintance with the subject it self. And those bredd to the use of musick, so as to have possession of it with eas and familiarity, have either bin mechanicks, or litle addicted to naturall inquisitions, or experiments. And to doe our author right, which none better then my self can, the subject never yet was touch't by a person more capable then himself, who was an excellent musitian both for hand, and composing, and not onely a good mathematician, and mechanick, but adicted to philosophicall inquirys and observations, wherein he had, as in all other thing[s,] a felicity of judgment incomperable.

But other greater affairs demanded his time, and this, with the rest of his ingenious essays, were but his stollen enterteinements;[16] and this in particular, is but a collection of hints, which it seems he was not minded to inlarge upon; or had not either time or pedantry enough to be a formall author. Therefore the notions, are but touch't, and being for the most part, and considering the intire hypothesis, so just, and according to truth, it is no wonder if they are capable of great enlargement and observation.

Page 7.

As to a tone, or single continuing sound, the deffinition of it is here truely given; and may otherwise be expres't to consist, in syncronous[17] pulses of the

---

[15] Marin Mersenne, *Harmonie universelle, contenant la theorie et la pratique de la musique*...(Paris, 1636) but more probably the Latin version, owned by RN, *Harmonicorum libri xii*...(Paris, 1635, 2d edn., enlarged, 1648). John Wallis, *Claudii Ptolemaei harmonicorum libri tres*...(Oxford, 1682). Isaac Vossius, *De poematum cantu et viribus rhythmi* (Oxford, 1673). For RN's brief comment on the last book, see *NP 2*, pp. 17–27; for the context, see my introduction in RN, *The Musicall Grammarian*, pp. 5–11.

[16] For these essays, collected in the 'Chapter of Ingenuitys', see RN, *Life/FN*, pp. 241–267.

[17] RN uses various forms of the word 'synchronous' to mean (*a*) equal timed, as here and at ff. 47, 47v, 48v, 51; (*b*) in equal spaces of time, ff. 47, 51; (*c*) equal spaces of time, f. 47; and (*d*) equality of time, ff. 48, 48v. For him, therefore, synchrony is synonymous with the concept, equality of timing—isochrony.

same body upon the air, in a quickness, whose intervalls are not distinguishable; and thus it is that the sence hath the idea of a continued sound, or ringing[18] as if it were a perpetuall emanation, and not consisting in distinct repeated effects. **[f. 44v]** So that the tone seems to be somewhat different from that which composeth it, as a clock is from brass, and this resulting or composed idea, is what wee are onely sensible of, without knowing the constituent matter and parts of which it is made; and therefore it seems wonderfull, but being ordinary wee pass it by, as wee doe earth, and stones, whose parts and composition wee doe not much inquire after.

Now to shew how our nature hath limited the capacity of our senses. Wee are able to distinguish what wee can number, or count. And wee can number, what wee can follow with the motion of any part of our body, either actually, or mentally by help of memory[s], which is the same thing, for wee cannot act in our minds things of which our bodys are uncapable. Therefore when individualls are to pass by in succession, if wee can tell[19] 'em, by nodding or [by moving] the finger, as is usuall in such occasion, so long wee distinguish them. But if they move so fast that wee cannot governe our parts, to note or move to each of them, the distinction of the individualls is lost, and they turne into the appearance of a continued stream, without interruption or distinction, which is a new idea, and different from the former.

For this reason it is, that a fire-stick waived, seems a stream of fire. Wee cannot make our ey fix upon it in any one position, but it must pass to another, and the memory of the former remaining, the whole represents a continued stream. Wee have from snow, reflected specks of light, such as every sphericall body being polish't, as water or ice is, sends out; but not being able to discerne the particulars, wee have a distinct idea, composed of the whole, which wee call white. The imperceptible starrs, when multitudes are together, by a joynt impression, appear in nebulæ. A caustick plaister, which macerates the severall parts that are touch't with it, produceth one sence, which is a painfull burning. And I will add one farther observa- **[f. 45]** tion, that when wee have inward paines, as the spleen,[20] and hypoconders,[21] so that wee doe not discerne the part affected, wee have a distinct thought, and fals opinion, that forrein things are

---

[18] *ringing*, i.e., resounding, resonant; hence, f. 45, a loud sound.

[19] *tell*: OED *v.* II. 21. *trans.* b. *spec.* to count heads (1511–1899).

[20] *spleen*: OED *sb.* 1.b. *obs.* the seat of melancholy or morose feelings (1390–1665). Here, the disease itself.

[21] *hypoconders*: OED *obs.* form of hypochondria *sb.* 1.a. and 1.b. as *pl.* of hypochondrium, the viscera, situated in the abdomen; the liver, gall-bladder, spleen, etc., formerly supposed to be the seat of melancholy (1547–1834). Here, the disease itself.

the caus of this paine.[22] So it is that malencholly persons are quasi madd, either upon religious, or worldly meditations. This production of a[n] idea of sound by a composition of pulses, will be remembred when wee come to speak of harmony, and chords,[23] where it is more elegant, and wonderfull.

It may seem strang[e] to some, why a tone should be so strong upon the sence, that is[,] so loud, when the pulses that compose it, are[24] so inconsiderable, as that of a small string or wyre, one stroke of which could not be heard. The instance of the nebulæ, answer[s] this.[25] For our senses, or attention, are not so quick, to catch every thing that occurs; there must be somewhat to draw and fix them, and then what was before obscure, shall be very manifest. As if you are told, a pin lys upon the ground, you shall look over it, and rays from it shall strikc your eye, 40 times, and still not see it, but when[26] once discerned and your eye levells a litle upon it, nothing is more manifest. But if a row of pinns were lay'd cross a room in a line, or many strowed about, you could not avoid seing them. When a single puls is made, if not of greater effect, then ordinary noises, it is gon, before your attention takes it, but if repeated continually, you cannot but observe it, and then it is very sensible, or loud.

So much as to the nature of tone, as it ariseth from indistinguishable pulses. Now as to the nature of the pulses; wee supose them to strike in equall spaces of time which produceth a grateful sound, or musicall tone. Otherwise the tone will be immusicall, and unpleasant, such as grating, whining, braying, shreeking, howling, and the like.

That this effect of even[n]ess, or unevenness, regularity or irregularity, upon our sence may appear reasonable, I will endeavour to examine the source of all pleasure, or disgust of that nature.[27]

[f. 45v]

It will be agreed to me that every thing, which is understood is pleasant, and that which is confused, and not understood is not so. For wee see, hear, and feel [in order] to know and distinguish things, which done, there is a satisfaction, or pleasure. But in the doing, there is a suspension, and upon disapointment, as when things pass, and new ones come, befor they are

---

[22] For his use of this observation, see RN, *Notes of Me*, pp. 201–202; for his irony, see Kassler, *Inner Music*, pp. 192–193.

[23] *harmony and chords*, i.e., FN's coincidence theory of harmony; see RN ff. 48v–50.

[24] MS has 'as'.

[25] See RN f. 44v, also ff. 42v–43.

[26] MS has 'whence'.

[27] For an expanded treatment of ff. 45–47v, see RN, *Cursory Notes of Musicke*, pp. 142–154; and RN, 'Cursory Notes of Building', *Of Building*, pp. 10–18.

knowne[,] there is a discontent or uneasyness. It is possible that many things very disorderly and unaccountable, may be not onely not ungratefull, but very pleasing to the sence; as landscapes of ruins, woods, cliffs, waters, meadows, etc., which doth not break our hypothesis, for such things have the advantage of our acquaintance with them, or with the like, and the consideration, of the historys, profits, uses, fertilitys and other things that belong to them, and therefore are injoyed with a prejudice. And it will not be denyed[.] But naturally, and abstracting our acquaintance with such things, as to an Hobbesian naturalist,[28] iff a landscape of such objects were reduc't to an order, and uniformity, one would pleas, and the other displeas. And wee answer ourselves in this when wee affect order and regularity, in trees, waters, buildings, and every thing els, when wee can procure it. And the offence from the contrary as well as the pleasure from that, are augmented by the comparison, and sensible difference found between the vertues of [the] one, and [those of the] other in the manner of touching our judgments. Tho I must needs say that life it self, being an emanation from the divinity, is *eo nomine*[29] pleasant, and life being sence, any objects are pleasant, rather then none, so [long as] they doe not dislocate our parts. And wee would rather see the most nauseous things, and hear the most ungratefull sounds then be shutt up in a dungeon, and rather not see nor hear at all. But when wee compare, and have choice, the differences of delight wee have, give the notions of pleasant to one, and the contrary to another, according to the degrees; and in such circumstances some are hatefull, and others, if I may so speak, incomperable.[30]

[f. 46]

Then taking it for granted that things understood are pleasant, I inferr that regularitys must be so, and irregularitys, the contrary. For from regularity, be it of what sort it will, wee have a notion, and it may be many, and comprehend the object fuller, then otherwise wee can doe. As when trees are planted in equall spaces, upon a strait line, there is one single notion of the whole; whereas if there were no designe in the planting of them, there were nothing but so many distinct objects, and not an object to the understanding from the aggregate, or method of disposing them.

---

[28] Perhaps RN alludes here to Hobbes, 'Elements of Philosophy', pp. 91–92, in which the author instructs the philosophical novice to imagine the world annihilated, because what would remain would be memories (past experiences) or, in RN's terms, 'acquaintances'. Taking the paragraph as a whole, the implication seems to be as follows. An observer without memories is uncultivated, so that objects reduced to an order would please or displease 'naturally'. But an observer with memories has knowledge by acquaintance, so that objects reduced to an order would please or displease by comparison.

[29] *eo nomine*: (Latin) of itself.

[30] *incomperable*, i.e., peerless and, thus, inexpressible (*je ne sais quoi*). For difficulties in anatomising beauty, see RN, *Cursory of Notes Musicke*, p. 206.

Wee have that service from our memorys, that wee can compare, and judg spaces without actuall measuring and lay one to the other, and say this is greater, duple, subduple, or equall with considerable accuracy, so that a great variation may be perceived, tho a small one is not, and therefore makes no difference.

Now when upon a view wee find, one part measuring another, so that there is a succession of discovery in the observation, and wee find, this here equall to that there, all level and upright, each side answering, parts of one measuring another[,] it cannot be denyed sure, that this knowledg springing from the order, and disposition of the whole doth administer more pleasure, then if every peice were pluck't from the other, and observed apart. And where there is not this symetry[,][31] wee have no faculty to know things by, or if the symetry be too fine and subtile.[32] For wee can measure one, with 3 or 4 but not with 100. And when an object hath not onely all the members, tho considered distinctly, in full perfection, but also forty times as many objects resulting from the symetricall economy of the whole, it must needs be forty times pleasanter then the members devided and uncomposed would make.

[f. 46v]

This is the truest account I can make, of the beauty of symetry, which is usually ascribed to an impression upon our nature, or a sort of devine stamp, or some such high supposition, without looking farther after it, whereas I take it onely to proceed out of the difference between knowing or not knowing, feeding the senses or starving them[,] which I thinck is a more sensible solution, then any miracolous in[ter]vention in our nature.[33]

---

[31] It is possible that the last clause in this sentence was added as an after-thought, but the meaning of the sentence would have been clarified had the clause been inserted before the editorial comma: 'And where there is not this symetry or if the symetry be too fine and subtile[,] wee have no faculty to know things by.'

[32] In the left-hand margin:

> This I speak as the effect of a criticall examen of the parts. And that I allow is the height of pleasure symmetry can afford. And an artist hath more pleasure then a cloune, but the less the examen is the less is the pleasure. But it must be some tho no act of comparison goeth along.
>
> For the figure is such as informes our spirit tho no voluntary intention goeth along with it[,] as a clowne shall thinck a fabrick beautyfull, which an artist judges of critically[;] and the symmetry gives us the same idea as if wee measured and compared it. So that this vertue of symmetry, works upon our sence, gratefully or hatefully, without our will accompanying. But if I may so say[,] insensibly.

[33] In 'Cursory Notes of Building', *Of Building*, p. 15, RN made this point more succinctly, writing that the beauty of symmetry is not 'in nature [i.e., innate], but [in] the experience of the materialls imployed, and the occasion and use of them, of which wee have a foreknowledge and thereupon calculate our approbation'.

And really and truely, there i[s] no naturall beauty or decorum in things, but it ariseth from our senc[e] and judgment, of which I have already spoke;[34] but I will [call][35] that an experience of usefullness, and the opinion of what is needfull. Too much, or too litle governes the decorum of most things, especially architecture which hath all its rules and proportions from thence.[36]

For nothing is beautyfull, and truely good in architecture, which is unreasonable. As to make solid over aperture, and contra, is absurd and hatefull, becaus the suport is broke, and how should the aperture support the solid? But solid upon solid, and aperture upon aperture is good, and fair, becaus usefull and proper. The *torre cadente* at Pisa, is a trick but not a beautiful fabrick.[37] Then why is an isoceles, more beautyfull then a scalene[?][38] Becaus in one [*Figure 4.B(i)*] the base hath a vniforme weight upon it, greater in the midle, and less at the sides, and the other [*Figure 4.B(ii)*] hath greater, and the greatest weight upon one point.[39] So in composing any order of 3[,] the midle should be greatest, and the sides[40] smaller, and lower [as in *Figure 4.C*], becaus the greater should have the advantage of abutment from them. And so for the height, width, length, and every proportion to be considered in building hath its test from the reason of us[e]fullness.[41]

To goe a step farther, the refined and establish't proportions assigned to the 5 orders of colum[ns] with their ornaments and entablature,[42] have be[en] insensibly derived from the same principle. **[f. 47]** For whence is it that wee

---

[34] See RN ff. 45v–46.

[35] MS is impossible to decipher here.

[36] Prior to c.1703 RN treated the subject of architecture in three texts: (1) 'Architecture', (GB:Lbl) Add MS 32540: ff. 69–80v, written after 1687 and before December 1691; (2) 'Of Building', (GB:Lbl) Add MS 32540: ff. 1–68v, written c.1695–96 but before RN's marriage in May 1696; and (3) 'Cursory Notes of Building', *Of Building*, pp. 1–103, written c.1698, representing RN's finished work on the subject. See Chapter 3, pp. 108–109 *supra*.

[37] In 'Cursory Notes of Building', *Of Building*, p. 18, RN described the leaning tower as 'an architectonick-whim, to make a tale for travelers but much unfitt for the purpose, a weighty fabrick, which should have no semblance of faling'.

[38] An isosceles triangle is a triangle with two equal sides (the third side is called the base and the angle opposite it the vertex). A scalene triangle is a triangle with no two sides equal.

[39] In 'Cursory Notes of Building', *Of Building*, p. 18, RN indicated that 'strength and firmeness is the true reason, why an equilater is more agreeable than a scalene tryangle'.

[40] MS has 'siders'.

[41] This is RN's second test of architectural beauty; his first test is 'derived from the understanding and comprehension of things'; see RN, 'Cursory Notes of Building', *Of Building*, pp. 12, 15 *et passim*.

[42] The five classical orders refer to column types: Doric, Ionic, Corinthian, Tuscan and Composite.

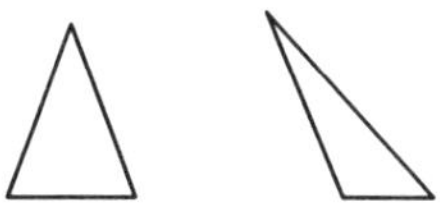

(i) (ii)

Triangles: (i) isoceles, (ii) scalene

Figure 4.B

Architectural order of three columns

Figure 4.C

say, this is too big for the height, or too high for the width, but from a generall knowledg of the strength of comon building-materialls and their ligaments[,] which tells us that less width, would have served, or more as necessary? That is, it is either stronger then needed, and then room is lost, or els it is not strong enough, and then the fabrick is weak and dangerous. So from the observation and practice of builders of best genius and skill, proportions have bin assigned between those extreams, which cannot be found fault with, and none will goe about it, when all the world will condemne 'em, for one man's caprice, is not a measure of excellence, but the agreement of many, and the best. But too much of this I fear.

Having shewed that order, and symmetry is pleasant with the reason of it,[43] I proceed to consider it in sounds. When pulses in sensible intervalls are in equall time wee plainely discerne it, and are angry when the order is broke by accelerating and retarding, becaus by what was past wee knew what is coming; but the new time breaks all our measures. The duration of time, and the length of space, is equally subject to the examination of our senses with the help of memory[s]. And in all wee doe wee affect equality and regularity in one as well as the other. Smiths, and carpenters usually strike in syncronismes; and it is usuall for workmen when they joyne, as smiths doe, (in great works especially) to fall into an order with which they are not a litle enterteined, and pretty mixtures of striking may be observed amongst anvill-and-anchor-smiths. In the idle shaking [of] our heels, or feet, and playing with our fingers,[44] wee fall into synchronisme and measure. And of noises, none is so insupportable as such of which wee have neither governement [n]or account, as the tolling of a funerall bell, or the clapping of doors and windoes by wind. This is apparent from our knowledg and experience in cases of sensible pulses[.]

Wee are to consider next of the insensible, or those whose intervalls are not sensible. **[f. 47v]** And the case will be found to be the same in those, as to the pleasing, or displeasing part. For tho wee have not a sence of each distinct intervall between, puls and puls, yet wee have a sence of the whole, which is compos'd of those parts. And becaus our bodys are so gross that wee cannot excite them to so quick an action as is required to distinguish those pulses, yet the nature of them is not changed; nor is our sence of them changed but both are the same. And wee may perceiv many things of which wee cannot give the anatomy.[45] Witness the instance of the splenetick old woman.[46] She is in paine,

---

[43] See RN ff. 45v (last paragraph)–47.

[44] MS has 'fingures'.

[45] *anatomy*: OED *sb.* III. 10. detailed examination, analysis (1569–1815).

[46] See RN f. 45, where the instance is to melancholy persons, not to 'the splenetick old woman'.

tho she cannot tell where it lys, and tho she calls it fear of want, or spirituall desertion or any thing els, yet the caus of her paine, and her sence (tho not opinion) of it is the same.

It must be granted that every variation in the action that toucheth our sence, varieth the perception, for sence is as minute, and infinite in capacity, as space and time are. So when there is a stream of continued pulses in equall spaces of time, the sence takes the idea of the spaces, as well as the quality[47] of the sound be it, bell, string, or pipe. And if it [the stream of pulses] varyeth from that order of spaces to any other, either quicker, slower, or incertein, the ear must also take it. And if the variation hath a relation or proportion to the former, or have no regard at all to it, the ear takes that too, and in short every variety of that kind that can bc. Then it follows,

1. That a tone, which changeth not, but continues the same now, anon, and till it ceaseth, must consist in syncronous pulses, for if they should slack or accellerate, the ear would find it.

2. That if the pulses are true, and uninterrupted, and unmixt, the idea of the tone will be such also; but if there be any thing rugged, or jagged in the tone (such as in the sound of a fals string, part of which vibrates slower, and part faster,) the falsness or inequality of the pulses is indicated; and the ey observing the even vibrations of a string (as the author elswhere observes[48]) discovers, whether the string will give a smooth, even tone, such as musick requires, or not.

[f. 48]

3. The synchronisme continuing, and the spaces of time changing from quicker, or slower, and contra, gives us the idea of what, in musick, wee call sharp and flatt; or high and low; treble and base. Or wee might as well call one a litle, and the other a great sound; and it would square as well with the idea[.] For the sharpning a tone, seems to contract it till it comes almost to a point or nothing but a squeek, and the flattning it seems to open it and make it wider, till the force is lost. Whereas in the nature of the thing, the difference is onely quicker and slower, in the space of striking. But if the strokes are not in equall time, call it what nois you pleas, it is no part of musick.

The compass of musick, is limited according to our senses. For a very great and heavy motion, such as of a rope, or huge cord, will vibrate so slow, that the pulses shall be sensible, as distinct, and then the idea of the tone, which is the result of imperceptible synchronisme in the pulses[,] is lost; besides[,] such sluggish motions, excite not a brisk compression in the aer (if any) to propagate such a motion as the ear shall take, unless it be very strong; and then

---

47 *quality*, i.e., secondary, subjective properties of sound, phenomena (here = tone quality or timbre); see also RN f. 51v (= pitch).

48 See FN p. 17.

it is but repeated dubbs such as a drum, or tub affords, but not what we understand by a tone. On the other side, when the pulses are so extream quick, as when they exceed the scale of musick, the effect becomes inconsiderable, and so at length, lost. Therefore the vsefull compass of musick is within the termes[49] of the ordinary scale which is exprest by organs, harpsicords, and violls, more then which either in flatt or sharp, is either impracticable or of litle or no use, which is what our author intends by his compass of musik.[50] And by [in]comensurate he means too great a distance[,] as 20, 40, 50, or 100, to one; which is quasi no proportion att all, in the use of our sence.[51]

And as to particular tones, as they are plac't in this musicall compass, the consideration of them will belong to another chapter,[52] for hitherto wee have not medled with any other then tones in generall.

[f. 48v]

Of the Relation, of Tones, and the Union of Mixt Sounds

This chapter[53] conteines the true and full mistery of musick, and is very plainely set forth, in a few words, which I goe not about to improve; and therefore shall not follow the paragraffs, but explaine the sence of the author more mechanically, and as I thinck more plausible to comon understandings, who are less apt to meditat[e], and, study. And without some paines to forme ideas, such as the author intended by his expression, the pleasure of his observations are lost, but on the other side, fully conceived must be extreamly pleasant.

And first of the creating a sound, for which wee will take the instance of a string as the most ordinary and examinable by sence.

The string sounding, moves visibly pendulum-wise from side to side, and so long the sound continues. And by the way it is to be noted, that the synchronisme of that motion, cannot be so mathematically proved, as the ear demonstrates, for if the motion were not such, the ear imediatly resents[54] it.

The sound is not made by the passage of the string, but at the instant of the returnes. For the string being in motion one way, carrys the ambient air with it, and upon the suddaine returne with a force, crossing that movement

---

49 *termes* = *terms*: OED *sb.* I. 1.a. *rare* or *arch.* limits, confines (13..–1855).

50 See FN p. 7.

51 See FN p. 8.

52 See f. 50.

53 See FN pp. 7–11.

54 *resents*: OED *v.* I.3. *obs.* feels or experiences (c.1650–80).

strikes a pressure upon the spring of the air, which dilates as was observed.[55] And this is the action which he calls vibration[56] and fills the air with litle circling force [i.e. waves], following successively as thick as the vibrations are. And if wee had any sence to survey it, wee should see the air full of circles, as the water would be, if any [driving] force should touch the calme surface, with syncronous pulses. Which would present a pretty sort of purle, especially if it had the varietys and mixture of circlings that wee shall speak of hereafter, in the progress of these discourses.[57]

[f. 49]

Now it is apparent, that the movement of this force in the air, must affect every thing in it, so as to impart some motion to [every thing in] it, more or less, as things are susceptible. And consequently every other string strained for sound must be moved also.

That which governes the celerity or dullness of the vibrations is the weight or substance, and the strength of the string, that is[,] the strength of its spring with regard to the weight of it, as the same string with a stronger tension shall vibrate faster, that is[,] more[58] acute, and a bigger string with the same tension shall be more grave and base.[59] So that it is possible, a string may be extended in such manner, that its vibrations shall be in the same spaces [of time], or in duple, triple, or any other proportions, with the other string before spoke of.

Now sett both strings a sounding together, if it shall fall out that they pass and returne together, the same way and att the same instants of time[,] it must be agreed, that the motion of the air raised by each shall promote reciprocally the vibrations of both. For the returnes will fall in the intervalls, and the motion, in the circle, which being the same way directed[60] must help it foreward. And this is what the author calls a coincidence of pulses;[61] which with unison strings happens in every returne. And if it should happen that at the beginning of the influence, these strings have upon each other, there should not

---

[55] See ff. 42–43v.

[56] See FN pp. 6, 11.

[57] Not in this text; but see (in chronological order) *NP 2*, pp. 35, 199–201; *NP 5*, pp. 47, 166–167; *NP 6*, pp. 27–35; 'The Theory of Sounds...', (GB:Lbl) Add MS 32534: ff. 8, 18 *et passim*; and *NP 3*, pp. 172–174.

[58] MS has 'the more'.

[59] RN gives a qualitative statement of the principles of vibrating strings: that there is a (quantitative) dependence of the frequency of vibration of a string on its length, tension and cross section (i.e., diameter and density). For the early history, see Dostrovsky, 'Vibration Theory', pp. 183–190.

[60] MS might read 'deflected'.

[61] See FN p. 8.

be so exact a conformity of effect, the impediment which they would be [to each other], would soon reduce the motion to it [resonance], and being there[,] will preserve it self. And this [balancing] operation cannot be the work of above 2 pulses, as the author observes,[62] therefore not perceptible to us.

On the other side if it should happen, that the spring of one is such that it will not vibrate in spaces of time, equall or in some other proportion **[f. 49v]** with the other, the force of each passing, must instead of helping the vibrations, be a mutuall impediment to each of them. For if a conformity happen by the coincidence of one stroke, the next cannot be so, and it may be not one in a 1000. Whereas in the other case, if there be a coincidence of any one stroke of unison strings[,] all the rest must be so.

The consequence of this is, [first,] the unison strings shall sound longer, then strings whose springs have no help from each other. And the more help each hath from the other, as when the coincidences are more frequent, and neer the vnison[,] the longer will the sound continue.

And secondly[,] this consent of movement must be easy and smooth to our ear, and the opposition must be grating, and uneasy, for the sence, according to our hypothesis[,][63] will be affected from the nature of the thing causing; and as to actuall pleasure and delight, that must arise out of a mixture and composition of various ingredients. And when wee speak of full consent, wee mean not most delightful, but onely free from clashing and offence. And when wee speak of harmony or musick in consort[,][64] which ought allwais to be actually pleasant to the ear, wee mean some vniforme and regular mixtures of pulses, in divers proportionate measures of time, as the ingenious seek to contrive when pleasant objects are to be presented. The palat is not taken with all consent[,] sweet or inoffensive, but a proportionate mixture of variety.[65] Onely this is to be lay'd downe as universally true[,] that sounds which have not frequent coincidences, but which[66] clash, an[d] oppose each other's motions, and are not kept *steddy* in any constant and respective movement, are ungratefull and hard, but others not so, and some actually delightfull, and in severall degrees according to the circumstances elswhere to be talked of.[67]

[f. 50]

---

[62] See FN p. 8, who says 'by the second or third pulse'.

[63] See RN ff. 44v–48.

[64] *in consort*: OED *sb.*[2] *obs.* in accord, in concert (1634–1793).

[65] In the left-hand margin: 'See after[,] *de gustibus n[on est] disputandum*': (Latin) there is no disputing about, no accounting for, taste. For 'after', see RN f. 50.

[66] MS has 'what'.

[67] See RN f. 50.

Pages 10. 11.

The matter of harmonious, or concording tones which wee call chords,[68] is well set out in this chapter and is the foundation of the arte of musick, tho few, or no artist (profest) understands it; I shall here onely take notice of the author's observation, that beyond the coincidence of 5 of the treble to 4 of the base, or beyond the 7th puls in either part is good, or concord; and he says it is becaus the coincidences are not enow to keep the motions steddy and regular to the most concording pulses.

I am apt to thinck there may be a steddyness kept in farther movements from unison, as the tone, or great second; which is a note wee can bear soundyng by it self a long time, so also the 7th in many cases. But out of musick, I confess those sounds are ungratefull. And it is doubdtless true that abstracted harmony is confined to those distances the author has made; but in composed harmony or descant, others are tollerable. But in some[,] as in places just (as wee call it) out of tune[,]it is not any where good.

And to be a litle farther criticall, I am apt to thinck, that even the worst of sounds, are kept steddy by each other; for wee find them allwais the same[—] as [the interval of] a tone, semitone, which is [the] worst[69] discord, is allwais of the same sort, and jarr [i.e., beat]. And as observed[70] the disorder of their motions shall produce in great pipes an alternation of sounding; as great semitone pipes in an organ beat like a drum. But when opposition is to a certein degree, the gust[71] doth not relish it, and of such matters no inquiry is to be had, more then I have given.

There is this reason to confirme what I say: that if discording sounds had not that effect upon each other, to maintaine some certein method of movements, whatever it is, but accident should take place; wee should observe that during the sound[,] variations would continually happen which doe not. Ergo, etc.

[f. 50v]

HOW TONES ARE PRODUCED, AND OF THE ASSISTANCES TO THE SOUND BY INSTRUMENTS

The forme[r] part of this subject I have spoke to as farr as strings are concerned,[72] except onely the breaking of their notes, which I reserve to the

---

[68] See FN pp. 3 and n.5, 7–11.

[69] MS has 'awors'.

[70] See FN p. 10.

[71] *gust*: OED *sb.*² *arch.* 1. the sense or faculty of taste (c.1430–1672).

[72] See ff. 44–45.

author's place.[73] As for the creation of sounds in wind instruments, his account is ingenious and new.[74] I shall differ with him onely in his systeme of an organ pipe, of that sort wee call the flute[,][75] that is, not the regoll,[76] but the comon flute or whistle pipe such as the flageolet, etc.

The author's account is that, the air eddys on the outside but not within, because of the protection there is from the sides of the pipe; therefore the filme [of air] having less support within then without, the eddy breaks thro by alternat forces, which makes the sound; as [in] the figure [see *Figure 4.D(i)*].

I find this is not true by tryall; for I have plac't an organ pipe of about a–re[77] comodiously; and when sounding, apply'd the flame of a candle to the orifice, so as the coal[78] of the week came below the vent[79] and nothing but pure flame touch't the orifice, and it was manifest that there was no breaking of the air thro the filme, for the flame was undisturbed unless upon the very palatt.[80] And there the flame seemed to splitt, and flutter both within and without. Holding the candle to the side of the pipe neer the orifice, the wind[81] would draw it [the flame] towards the midle and there split and flutter as before. This is all I could discover by the eye and stands represented in the figure [see *Figure 4.D(ii)*].

Then to consider the true reason of the sounding of this sort of pipe, wee must lay aside the former supposed action[—]which by the way was not [un]reasonable to imagin, that an eddy unpent, as in the outward air[,] should break thro the principall stream, from which its force is derived secondarily, and therefore [should] weaken[—]and seek some other action, analagous to the more visible actions that produce sound.

And that is[,] as I conceiv, the filme of air falling upon the pallat or opposite edg, with an indifference **[f. 51]** or equall tendency, outward, or inward. It crosseth the pallat in an alternate motion, and sometimes strikes all

---

[73] See FN pp. 17–21.

[74] See FN pp. 11–15.

[75] *flute*, i.e., flue pipe; see Figures 2.1 and 3.3 *supra*.

[76] *regoll*, i.e., reed pipe.

[77] *a-re*, i.e., 'a' is the pitch, 're' is the solmisation syllable for that pitch.

[78] *coal*: OED *sb*. I. 1.c. *obs*. the glowing portion of a match (1590). Here, the lighted portion of the 'week' (*obs*. form of wick).

[79] RN uses the term 'vent' to mean: *(a)* the slit between the lower lip of the pipe and the languid, as here; *(b)* outlet, f. 51v; and *(c)* side hole from which the sound mostly radiates in a pipe closed at one end, f. 51v.

[80] *palatt*, i.e., edge; see also RN f. 51.

[81] *wind*, i.e., air coming from the slit.

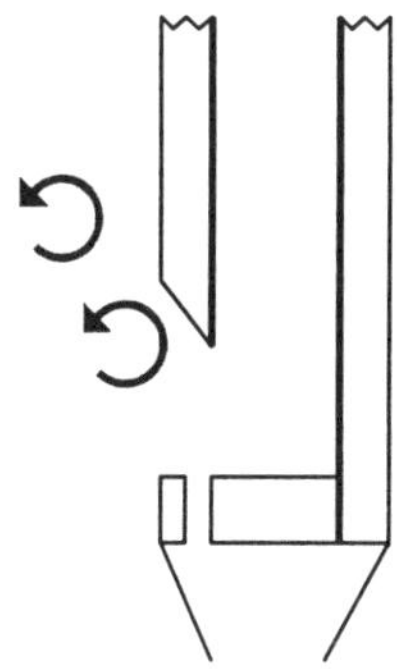

(i)

(ii)

Flue pipe as driven by edge tones
(i) Francis North's theory
(ii) Roger North's experiment refuting that theory

Figure 4.D

into the pipe and sometimes all out of it; and every returne, is a syncronous stroke upon the included air of the pipe; which must be a tone. To prove this:

1. A filme of air being thin, and of equall stren[g]th is in nature of spring, (without compressure) and may be bent by an outward force either way, and returne. Wee see the same in a stream of water which is capable of no compressure; as when a ball is put in and supported by a *ject d'eau*;[82] the weight of it shall carry the stream one way and another synchronously, as a pendulum erected. Wherefore the filme of air is capable of the like action, when a regular [driving] force toucheth it.

2. And as for that it is to be considered; that when that thin filme of air enters the pipe, or suppose but one part of it, the whole cilinder of air is affected to move, but as a greater body will not be moved with a celerity equall to that of a less which strikes it, so the stream entring shall fill the cavity and make a compressure at the mouth, before there is any at the upper end; and it must come there in time. This compressure bears upon the filme as it doth upon all the sides of the pipe thereabouts and determines the indifference; so that the whole filme shall bear out of the pipe. And it is to be noted that there is no compressure without, to bear the filme into the pipe, becaus there is no restraint but the whole air free to expand in.

This[83] is the action that bears the filme out of the pipe[;] there will be found the like to bring it back. For being in motion, the filme draws the continuous air with it and makes a laterall stream neer the surface by its parts laying hold of others, and consequently must draw air out of the pipe. And by the same cours as before, instead of a compression produces a dilatation of the air within the orifice[,] which makes way for the atmosphear to thrust it back againe.

So[84] that this alternate compressure, and delatation of the air within neer the orifice of the pipe, must make the filme play, [a]cross the pallat perpetually, if it be so indifferently plac't that the force be not one way too strong for it.

[f. 51v]

Now to give the mechanicall reason why the air in the pipe governes the pulses at the mouth, or the vibrations of the filme[,] wee must first allow these principles, viz.,

---

[82] *ject d'eau*: (French, *obs.* = jet d'eau) a water jet or fountain.

[83] In the left-hand margin:
Besides, the *vis impressa* outwards shall carry it too farr, as the swing of a pendulum, and by its proper direction, be brought back, beyond the strait cours.

[84] In the left-hand margin:
If an hole be made in the pipe neer the mouth that letts out the air, and prevents the compressur[e] the pipe will not speak at all, which demonstrates this hypothesis.

1.[85] That the freer passage the air hath in the body or cavity of the pipe, the quicker the pulses will be; [and] the smaller the body be, the quicker[,] etc.[86] For the compressure filling the cavity, drives out and tends to keep out the filme[;] onely when it is spent and a dilatation succeeds, it returnes. Therefore the sooner the compressure is spent,[87] the sooner the returne is, and in an open vented pipe[88] or [one] which hath a small quantity of air, or a small pipe, the compressure is spent sooner then in a greater. Therefore a short colume of air in a pipe, shall make the sound more treble then a longer [column].

I may add to confirme this hypothesis that, [as] the compressure in such cases is sooner effected, as well as withdrawn, so on both sides it conduceth to accellerate the pulses.

From hence it is that organ makers find that a close[d] pipe will be at least an octave more base then the same pipe open.[89] For the air hath no vent but at the mouth during the intervals of filling and therefore is of tender sound; and to eas such pipes, they use to give a small vent, at the end,[90] and so draw off much of the labour from the mouth. The tuning of these pipes is by lengthning and shortning the cavity; for if it be made longer there is more cavity to fill, and if shorter less. And open pipes are tuned by bearing a peice of mettall over the open end, which hindring the issue of air there more or less, produceth the effect proposed.[91]

Now it is to be considered, that if a pipe be longer then a due proportion, the passage of the air will be hindred by the sides, as all contact is an impediment to motion, and then the pipe asumeth a quality like one that hath less air, in a freer cavity, and shall break into a more acute tone; that is[,] the

---

[85] In the left-hand margin:
Query short close[d] pipes

[86] The part of this sentence after the semi-colon may have been a later insertion, since it is written, indented, above the sentence beginning, 'For the compressure'.

[87] In the left-hand margin:
This caveat

[88] *open vented pipe*, i.e., an open diapason pipe, in which the sound is made to radiate partly from the open end and partly from the side hole.

[89] This is true if the pipes are the same length.

[90] The hole is at the side, not at the end.

[91] In the closed pipe, the upper end is fully closed by a plug, the adjustment of the precise position of which provides the tuning mechanism. In the open pipe, the nature of the tuning mechanism depends on whether the pipe is made of wood or metal. RN describes the tuning mechanism for wood pipes, where the effective length is controlled for tuning purposes by a slot running down a short distance from the top that can be partly covered or uncovered by a bent strip of metal.

vibrations of the filme will be more quick. And this shall happen or not, in a nice case, as the blast is stronger or weaker. For upon a gentle blast[,] the air hath time to pass, and governes the vibrations in the ordinary way, but if the wind presseth faster upon the pipe **[f. 52]** then it [the air] can pass consistent with those measures,[...][92]

92 On f. 52v the words 'on the Essay of Musick' are written sideways in the right-hand margin, but the rest of the page is blank.

APPENDIX B: SOME 'INGENIOUS ACQUAINTANCES' OF FRANCIS NORTH

Since the friend to whom Francis North directed his *Essay* remains an unsolved problem, the purpose of this appendix is to identify those persons whom Roger North named as his brother's 'ingenious acquaintances'—ingenious, because they shared an interest in philosophy, music or other arts.[1] Roger North is still the principal source for biographical details about his brother—the person referred to as 'his Lordship' in the passages quoted below. But this epithet does not necessarily mean that the 'ingenious acquaintances' were formed after 1683, when Francis North was raised to the peerage. The private papers of Francis North have yet to be thoroughly investigated, including all the letters to and from Francis North for the period prior to 1677. It seems, however, that the letters held by the British Library, London, the Bodleian Library, Oxford, and the Huntington Library, San Marino, California, concern either family, estate and financial matters or law cases, not matters relating to the modern philosophy of music.

Aubrey, John (1626–97)

> There was one Mr. Aubrey of Surrey, a profest vertuoso, and allwais repleat with new discoverys, he often visited his Lordship, who encouraged him by his attention, and asking many questions, and his answers served well enough, in order to a melius inquirendum [better examination].[2]

On 6 April 1646, Aubrey was admitted to the Middle Temple, London, where he was 'off and on'; but he was never called to the bar. In 1663 he was elected a fellow of the Royal Society, afterwards travelling abroad. Upon his return, he wrote a sketch of his 'Idea of education of young gentlemen' (1669), which he commenced rewriting in the 1680s and completed about 1683. In a letter to Elias Ashmole, dated 22 July 1684, he mentioned visiting Francis North.[3]

---

[1] RN, *Life/FN*, pp. 133, 242–252; for other avenues of investigation, see also *NP 4*, pp. 1–46.

[2] RN, *Life/FN*, p. 243. For a survey of Aubrey's life in the context of the times, see the editor's introduction, Dick (ed.), *Aubrey's Brief Lives*, pp. 17–160.

[3] Josten (ed.), *Elias Ashmole*, vol. 4, p. 1739; for FN, see also vol. 4, pp. 1607–1608, 1716, 1792, 1797. For a possible reference to DN (1), see vol. 4, pp. 1840–1841.

Because of financial difficulties, he often changed his address; and his lodgings included the residence of his patron, Edmund Wyld (q.v.) and the rooms of his friend, Robert Hooke. His own circle of friends and acquaintances, which was very large, included, besides those just mentioned, John Evelyn (q.v.), Thomas Hobbes,[4] John Hoskins (q.v.) and Jonas Moore (q.v.). On 10 December 1675 he joined Hooke's 'New Philosophical Club'.[5] In his 'Idea', Aubrey mentioned Francis North several times.[6] He also asserted that music is reducible to 'the laws of geometry', perhaps echoing Hobbes's sentiments;[7] but he seems not to have understood Hobbes's way of applying geometry to physics, where imaginary 'lines' are conceived as elastic and, hence, as having 'powers'.[8]

Ball [Balle], William (d. 1690)

> His Lordship had another vertuoso acquaintance in the [Middle] Temple Mr William Ball, [eldest] son of Sir Peter Ball, that from his first society of that kind kept him company, but family misfortunes overtaking him, his activity, and inclinations hebetated, or rather degenerated into domestique cares, so that I think it enough to have named him, as one in the list of [his] Lordship's ingenious acquaintance.[9]

According to Aubrey, the 'first beginning of the Royal Society (where they putt discourse in paper and brought it to use) was in the Chamber of William Ball ... in the Middle Temple'.[10] At this inaugural meeting, held on 28 November 1660, it was decided that weekly meetings would convene on Wednesdays at rooms in Gresham College and in the vacations in Ball's room at the Middle Temple, the latter decision being rescinded in the following month.[11] Ball's name, with that of his brother, Peter, a physician, appears in the Society's first charter of 15 July 1662; and he was the Society's first treasurer, resigning on 30 November 1663. His interest in the propagation of motion was recorded in the Society's records in 1675, when Ball conducted experiments

---

[4] In 1670 Aubrey presented a portrait of Hobbes to the Royal Society; see Hoppen, *The Common Scientist*, p. 242.

[5] Hooke, *The Diary*, p. 199; see also pp. 200, 202, 205–207, 208.

[6] See Stephens (ed.), *Aubrey on Education*, pp. 28, 37, 56, 163 n.6, 165 n.5.

[7] Stephens (ed.), *Aubrey on Education*, p. 113.

[8] See Kassler, *Inner Music*, p. 103.

[9] RN, *Life/FN*, p. 244.

[10] Dick (ed.), *Aubrey's Brief Lives*, p. 480; see also p. 67.

[11] Purver, *The Royal Society*, pp. 131–133.

with Christian Huygens.[12] Aubrey reported that Ball had 'a good Copie' of Hobbes's portrait, drawn shortly before 1679, so he may have been one of Hobbes's 'ingeniose and learned Friends'.[13] Between 1679 and 1680 Ball resided in Devonshire, where he carried on astronomical work and corresponded with Robert Hooke.[14] But he died in London and was buried in the Round of the Middle Temple.

Evelyn, John (1620–1706)

> His Lordship once was invited to a philosoficall meall, at the hous of Mr Evelin at Deptford. The hous was low, but elegantly set off with ornaments, and quaint mottos at most turnes, but above all his garden was exquisite, being most boscaresque and as it were an exemplar of his book of forrest trees. They appeared all so thriving and clean, that in so much variety, no one could be satiated in viewing, and to these were added plenty of ingenious discourses, which made the time short.[15]

Evelyn was admitted to the Middle Temple, London (1637), where, after travelling abroad, he took chambers. Upon establishment of the Royal Society by royal charter on 15 July 1662, he was elected a council member and, in 1672, secretary. In that year Francis North 'specially admitted' Evelyn's son to the Middle Temple.[16] Evelyn, whose diary contains a number of references to members of the North family, describes Francis North as 'a most knowing, learned, ingenious gent, & besides an excellent person, of an ingenuous sweete disposition, very skillful in Musick, Painting, the new Philosophie & politer studies'.[17]

Flamsteed, John (1646–1719)

> Sir Jonas [Moore] once invited his Lordship to dine with him at the Tower, and after dinner presented Mr Flamstead; his Lordship received him with much familiarity, and encouraged him to come, and see him often; that he might have the pleasure of his conversation; the starr-gazer

---

12 Birch, *The History of the Royal Society*, vol. 2, p. 132.

13 Buchanan-Brown (ed.), *John Aubrey*, p. 452.

14 Birch, *The History of the Royal Society*, vol. 3, p. 517, and Hooke, *The Diary*, pp. 440, 452.

15 RN, *Life/FN*, p. 243.

16 Evelyn, *The Diary*, vol. 3, pp. 613–614.

17 Evelyn, *The Diary*, vol. 4, p. 299; see also vol. 4, pp. 87, 97, 312, 340, 361, 365, 372, 383, 393, 445, 466, 546, and vol. 5, pp. 309, 317, 365, 366, 529, 531.

> was not wanting to himself in that, and his Lordship was extreamly delighted with his accounts, and observations about the planets, especially those attendant on Jupiter; shewing how the eclipses of them, being regular, and calculable, might rectifie the longitudes of places upon the globe; and demonstrated that light passed not in eodem instanti, but in time; with other remarkables in the heavens; which discourse allwais regaled his Lordship; and a good benefice falling void, not farr from the observatory, in the guift of the Great Seal, his Lordship gave it to Mr Flamsted, which set him at eas in his fortunes, and encouraged his future labours, from which great things were expected; as applying the Joviall obser[va]tions to marine uses, for finding longitudes at sea, and to correct the globes celestiall, and terrestriall; which were very faulty. And in order to the first, he had composed tables of the ecclipses of the satellites, which shewed when they were to happen, one after another; and of these, finely pasted upon neat bo[a]rd, he made a present to his Lordship. And he had advanced his other designe of rectifying mapps, by having provided large blank globes, on which he might inscribe his places corrected, but plenty, and paines seldome dwell together, but as one enters, the other gives way; and in this instance a good living[,] pensions, etc., spoyled a good cosmografer, and astronomer, for very litle is left of Mr Framstead's [sic] sedulous and judicious applications that way.[18]

Flamsteed, who spent the first twenty-five years of his life in Derby, came to London about 1670, when he was introduced to Jonas Moore (q.v.). On the invitation of Moore, Flamsteed removed to London in 1674, residing with Moore in the Tower. Shortly afterwards, Moore introduced him to Francis North.[19] Due to Moore's intervention, Flamsteed was appointed first astronomer royal on 4 March 1675, when he moved to Greenwich; in the same year he took orders. On 8 February 1677 he was elected a fellow of the Royal Society; and in 1684 North granted him the living in Burstow, Surrey.[20]

Hoskins [Hoskyns], John (1634–1705)

> His Lordship had great pleasure in the society of Sir John Hoskins, who was bred towards the law, but a good estate, and strong inclination to the philosoficks, got the better of that designe. ... The good knight was for

---

[18] RN, *Life/FN*, pp. 243–244.

[19] In a letter to Moore, dated 5 January 1676, Flamsteed requested that his paper on the celebration of Easter be passed on to FN; see Flamsteed, *The Correspondence*, vol. 1, p. 405; see also vol. 1, p. 726 and vol. 2 p. 266.

[20] Baily, *An Account of the Revd. John Flamsteed*, pp. xxviii–xxix. For an assessment of his work, see Thoren, 'John Flamsteed'.

> many years, president of the Royall Society, and I have given him an advanc't post among his Lordship's contemporarys.[21]

A barrister in the Middle Temple, master in Chancery and M.P. for Herefordshire (1685), Hoskins was active in the affairs of the Royal Society, of which he was an original fellow (1663), president (1682–83) and secretary (1685–87). When Theodore de Vaux, another fellow (elected 1665) and physician to Charles II, invited Francis North to become a fellow, it probably was at the instigation of Hoskins, who on 26 February 1674 was sworn a member of the Society's council, having previously been present in a legal capacity.[22] In the following year Hoskins joined Robert Hooke's 'New Philosophicall Club' established on 10 December 1675.[23] In 1680 he succeeded his father in the baronetcy, having been knighted sometime previously.

Lely, Peter (1618–80)

> His Lordship allwais loved to see variety of good pictures, but was a minor crittique, 'till his acquaintance, and association with Sir P. Lely was embrased; and then he began to see, how litle he understood of the matter.[24]

Born in Westfalia, he came to London, where on 26 October 1647 he was made a freeman of the Painter-Stainers Company and could take apprentices. In 1648 one of his apprentices was Robert Hooke, who did not stay long, because he was allergic to paint. Lely entertained at his house on the Piazza in Covent Garden. His neighbour was the 'limner', Samuel Cooper,[25] and his circle of friends included Hugh May (q.v.) and Samuel Morland (q.v.). In 1680 he was knighted. Like his colleague, Grinling Gibbons, Lely's musical interests manifested themselves in his art.[26] But he was best known as a portrait painter; and on 13 August 1668, Francis North wrote to his father that 'Tomorrow I

---

[21] RN, *Life/FN*, p. 242; see also Chapter 1, pp. 35–36 *supra*.

[22] See RN, *Life/FN*, p. 242 (FN declined the invitation to become a fellow).

[23] See p. 202 n.5 *supra*.

[24] RN, *Life/FN*, p. 251; see also pp. 45, 57, 133, 196, 252, 477.

[25] A friend of Aubrey, Cooper instructed Hooke in drawing; he also was acquainted with Hobbes and drew his portrait. See Buchanan-Brown (ed.), *John Aubrey*, pp. 394, 431–432, 449.

[26] See Millar, *Sir Peter Lely*, pp. 10, 11, 40, 41, 42, 48. For Lely and Gibbons, see Esterly, *Grinling Gibbons*, pp. 23, 55–59, 206–208, 218.

shall sit (I hope the last) for my Picture by Mr. Lillye'.[27] Since Lely was averse to business, he required help in settling his affairs, especially after the death of his wife in January 1674.[28] Francis North provided the necessary assistance (including appointing Roger North as a trustee); and Lely returned the favour by painting portraits of some of North's relations.[29] Lely himself had amassed a magnificent collection of pictures, including drawings by old masters. As one of the executors (by Lely's will dated 4 February 1680), Roger North catalogued the collection for auction,[30] which took place in 1688.[31]

May, Hugh (1622–84)

> ...his Lordship...was covetous of nothing more then the society of the vertuosos of his time, as Lely, Moreland, May, Moor, Flamstead, and others of that tribe, who all courted him, and embraced his conversation, and many owned to profit by his encouragement, and protection.[32]

Courtier, architect and cousin of Baptist May, with whom he is often confused, Hugh May spent part of the Commonwealth period in the employ of George Villiers, 2d duke of Buckingham, living for an extended period in Holland. For a short time, he seems to have worked in the studio of Peter Lely (q.v.), with whom he resided in the 1650s. In 1660 he was made paymaster and, in 1668, comptroller of the king's works, a post he held until his death. After the great fire, he was one of the commissioners for the rebuilding of London.[33] In 1673 May was appointed comptroller of the works at Windsor Castle, the rebuilding

---

[27] (GB:Ob) MS North c.4: f. 228; see also f. 245 ('I sent my Picture down last weeke', dated 4 March 1669) and f. 315 ('I am now going to Mr Lilly to sit for my <...> wch I hope will be finished to be sent to Catlidge', undated). Peter Millard in RN, *Notes of Me*, p. 32 and n.103, indicates that, after FN's death, RN took possession of this portrait; see also RN, *Life/FN*, pp. 196, 252, who states that Lely also painted a half length of FN and his wife.

[28] Millar, *Sir Peter Lely*, p. 22.

[29] See RN, *Life/FN*, pp. 196–198. For the role of FN's brother, DN (2), see Grassby, *The English Gentleman in Trade*, pp. 69, 83.

[30] See Peter Millard in RN, *Notes of Me*, pp. 318–320; see also Millar, *Sir Peter Lely*, p. 24.

[31] See Detholoff, 'The Executors' Account Book'; for the account book itself, see (GB:Lbl) Add MS 16174, dated 1679–91.

[32] RN, *Life/FN*, p. 133.

[33] See Reddaway, *The Rebuilding of London*, pp. 55, 108 n.2, 312. As a city surveyor, Hooke would have come into contact with May; see Hooke, *The Diary*, pp. 38, 413.

of which continued until about 1682.[34] He and Lely were patrons of Grinling Gibbons; and in Lely's self portrait with Hugh May, Gibbons stands behind the two men.[35] At Lely's death, May, Roger North and the physician, William Stokeham were appointed executors of Lely's estate.[36]

Moore [More], Jonas (1617–79)

> His Lordship had a great value for Sir Jonas Moor, a capitall mathematitian, knowing well his worth and honesty, by means of his imployment under the commission for dividing the fenns; at which his Lordship had presided. That good man had taken Mr Flamstead, the noted astronomer, into his protection; and when he was scarce able to subsist in his colledg at Cambridg, planted him in the Tower, with accomodation in the buildings of the ordonnance, of which Sir Jonas was an officer; procured him instruments, and at last setled him in the new built observatory at Greenwich. Sir Jonas once invited his Lordship to dine with him at the Tower, and after dinner presented Mr Flamstead....[37]

In 1647 Moore was mathematical tutor to the duke of York and in 1649, through the patronage of Edmund Wyld (q.v.), surveyor-general of the Norfolk Fens drainage system.[38] At first he lived wherever needed; but in 1652 he moved to a market town, and in 1653 to Southery, Norfolk, where he stayed until leaving the Fens in 1657. From 1649 Francis North's father, Dudley, was a Fens commissioner,[39] so that Roger North seems mistaken in stating that his brother, Francis, presided over the Fens commission at that early date.[40] In 1663 Moore travelled to Tangier; upon his return to London he occupied various addresses until his appointment as surveyor-general of the ordnance (1663). From thence he resided in the Tower. He was knighted in 1673 and elected a fellow of the Royal Society in 1674, the year he invited John Flamsteed (q.v.) to London. A collection of his papers on applied mathematics

---

[34] See Reddaway, *The Rebuilding of London*, pp. 51 n.1, 57–59 and n.3, 71, 107, 312; see also Esterly, *Grinling Gibbons*, pp. 33, 56–57, 64, 66, 206, 218.

[35] Reproduced in Esterly, *Grinling Gibbons*, p. 57, and (less clearly) in Millar, *Sir Peter Lely*, p. 67. For the patronage of Robert Gayer, RN's future father-in-law, see Esterly, *Grinling Gibbons*, pp. 101, 215–216, 220.

[36] RN, *Notes of Me*, pp. 238, 249, 318–320.

[37] RN, *Life/FN*, p. 243.

[38] See Willmoth, *Sir Jonas Moore*, pp. 88–120.

[39] See Abbreviations, p. xi *supra*; see also Randall, *Gentle Flame*, pp. 58–59, 72.

[40] On 19 September 1676 Robert Cotton was appointed a commissioner in place of FN, who was 'incapacitated by being interested' (GB:Lpro) C226/4/16 (28 Chas II).

was published as *A Mathematical Compendium* (London, 1674) with a dedication to John Werden (q.v.). One of the papers treats applications of the pendulum to clockwork.[41] During 1675 Moore became involved with Robert Hooke's attempt to obtain a patent for a spring-balanced watch;[42] and on 10 December of that year he joined Hooke's 'New Philosophicall Club'.[43] According to Aubrey, Moore 'came to the Middle Temple London', where he published his first book, *Moore's Arithmetick: Discovering the Secrets of that Art* (London, 1650).[44] And when Moore was surveyor-general of the ordnance, he told Aubrey of his 'great veneration for' Thomas Hobbes.[45]

Morland [Moreland], Samuel (1625–95)

> About this time [1675], the philosoficall world was enterteined about setling the grand affair, of the mercuriall barometer and its indications. Among the rest Sir Sam. Moreland published a sheet conteining a device to prolong the indicatory space, from 3 inches, as in comon tubes, to a foot, or more ad libitum; he defyed all the vertuosi to resolve it and this he called a statick barometer. ...his Lordship wrote a paper in answer to the knight's challenge, and considered this experiment according to the laws of hydrostaticks.... When the vertuoso received his Lordship's paper sent him for answer he blustered and threatened a most powerfull answer but never was so kind to send any; and on the contrary took an opportunity to wait on his Lordship, and they became good philosoficall freinds and acquaintance[s]....[46]

Morland, a tutor of Samuel Pepys,[47] was knighted in 1660.[48] He is known chiefly for his inventions, some of which were patented. These include a perpetual almanac, two calculating machines, a speaking trumpet[49] and the

[41] The topic must have been discussed with Humphry Adamson, who, in 1673 was living in the Tower with Moore and was the tradesman who made Huygens' cycloidal cheeks for pendulums; see Dickinson, *Sir Samuel Morland*, p. 31, and Willmoth, *Sir Jonas Moore*, p. 149.

[42] See Willmoth, *Sir Jonas Moore*, pp. 175–176.

[43] See p. 202 n.5 *supra*. For Moore's death and funeral, see Hooke, *The Diary*, pp. 422–423.

[44] Buchanan-Brown (ed.), *John Aubrey*, p. 212.

[45] Buchanan-Brown (ed.), *John Aubrey*, p. 450.

[46] RN, *Life/FN*, pp. 245–246.

[47] Pepys, *Private Correspondence*, vol. 1, p. 111.

[48] For a 'Conspectus Vitae', see Dickinson, *Sir Samuel Morland*, pp. xiv–xv.

[49] For details of the speaking trumpet, see Dickinson, *Sir Samuel Morland*, pp. xv, 40–44, 109 and Plate 5. For RN's first reference to this invention and the pamphlet that describes it, see *Cursory Notes of Musicke*, pp. 99–100.

balance ('static') barometer. He also devised mechanical methods of raising water and made improvements in fire engines. John Playford published Morland's perpetual almanac as *Vade mecum or the Necessary Companion*...(London, 1679). Simon Beale, one of the king's trumpeters, manufactured his speaking trumpet;[50] and Humphry Adamson, a tradesman who, in 1673, lived in the Tower with Jonas Moore (q.v.), made his calculating machines. Roger North's statements, quoted above, can be dated with some accuracy. As mentioned previously,[51] he and his brother, Francis, had a dispute concerning the barometer, after which they each sent their papers to the Royal Society's secretary, Henry Oldenburg, who died in September 1677. In that year Robert Hooke devised a prototype wheel barometer, some parts of which were utilised in the balance barometer invented by Morland.[52] In 1678 Morland published a sheet concerning this instrument, in which he denied Boyle's hydrostatic doctrine that the mercury in the barometric tube was supported by the pressure of the atmosphere. Instead, he promoted the hypothesis of Francis Line and his supporter, Matthew Hale. Shortly afterwards, Francis North sent Morland a paper, in which he argued the case for Boyle's doctrine. Subsequently, North introduced the barometer as an item of commerce by instructing two instrument makers, Henry Jones and Henry Wynne, how to manufacture them.[53] Since North and Morland became acquainted after publication of the *Essay*, this would eliminate Morland's name from the list of candidates for the friend to whom the *Essay* was directed.

Werden [Worden], John (1640–1716)

> His Lordship had another acquaintance, who resided in the [Middle] Temple, and being of a retired disposition was very farr gone in the misterys of algebra, and mechanicks. This was Mr John Werden,

---

[50] See Ashbee and Lasocki, *A Biographical Dictionary of English Court Musicians*, vol. 1, pp. 135–136.

[51] See Chapter 1, p. 27 *supra*.

[52] Hooke, *The Diary*, p. 336, first mentions Morland's barometer on 24 December 1677.

[53] For Wynne [Wynn, Win], see Hooke, *The Diary*, p. 343 *et passim* and Brown, 'Guild Organisation and the Instrument-making Trade'. For FN, Wynne and Morland, see RN, *Life/FN*, pp. 245–248, where RN records that Wynne also made a 'case' of mathematical instruments for FN. Bryden, 'Sir Samuel Morland's Account', fills in the gaps in Middleton, *The History of the Barometer*, pp. 82, 100–103. Subsequent to FN's paper, RN produced the first of his three discourses on the barometer: (1) 'An Essay concerning the reason and use of the Baroscope', c.1680s, (GB:Lbl) Add MS 32543: ff. 23–33v; (2) 'An Essay concerning the Reason & use of ye baroscope', c.1690s, (GB:Lbl) Add MS 32543: ff. 1–22v; and (3) 'The Air-Gager ... of the Barometer...', begun c.1713, completed c.1716, (GB:Lbl) Add MS 32542: ff. 1–114v.

> afterwards Sir John, and many years a commissioner of the Customs. He was the onely son of an incomparable courtier, Cavalier, and a most faithfull servant in the royal family, Collonel [Robert] Werden. This Sir John inspired his Lordship with a sort of fury, in persuit of the art of perspective; he shewed him the picture of a tree, upon the bows of which hung the letters of his name cut solid, and placed as it were contingently, but expressed in true perspective.[54]

In 1660 Werden was called to the bar at the Middle Temple. Between c.1665 and 1672 he was employed in diplomatic service; and in 1672 he was created baronet. In 1674 the *Mathematical Compendium* of Jonas Moore (q.v.) was published with a dedication to Werden, in which he is described as secretary to James, duke of York.[55] Subsequently, he was a member of the queen's council (1686). As one of the six commissioners of customs (1685–97, 1702–14), Werden came into contact with, and taught algebra to, Francis North's brother, Dudley.[56]

Wyld [Wild], Edmund (*fl.* 1649–93)

> One Mr Weld [sic], a rich philosopher lived in Bloomsbury; he was single, and his hous a sort of knick-knack-atory, most of the ingeniosos about towne sometimes dined with him, and among the rest his Lordship did his suit and service there. This gentleman was of a superior order, and valued himself upon new inventions of his owne. He sowed sallads in the morning, to be cutt for dinner, and claimed the invention of painted curtaines in vernish upon silk which would bend, and not crack and his hous was furnished with them; and he delighted in nothing more then in shewing his multifarious contrivances.[57]

Wyld, an original fellow of the Royal Society (1663), was described by Samuel Hartlib as a Commonwealth M.P. 'and of the Tempel[,] a meere Naturalist but a great Lover and Advancer of Sense and Experim[ental] Philos[ophy]'.[58] He was a patron of John Aubrey (q.v.)—who seems to have regarded him as a

---

[54] RN, *Life/FN*, p. 243; see also pp. 197, 477.

[55] Between 1673–75 Werden is mentioned several times in Hooke, *The Diary*, pp. 54, (possibly 134), 163, 169, 176, some memoranda of which suggest that, like Moore, he was involved with Hooke's attempt to patent a spring-balanced watch.

[56] Grassby, *The English Gentlemen in Trade*, p. 196.

[57] RN, *Life/FN*, p. 243.

[58] See Willmoth, *Sir Jonas Moore*, pp. 105–106; see also pp. 19, 42–43, 102, 143, 176. Hartlib's various descriptions of Wyld date from 1649, whereas Hooke, 'The Diary', vol. 10, p. 253, mentions meeting Wyld in 1693.

father figure, Jonas Moore (q.v.)—who taught him mathematics, and William Petty.[59] On 10 December 1675 he joined Robert Hooke's 'New Philosophicall Club'.[60] Aubrey, from information supplied by Francis North, reported that Wyld was acquainted with 'Captain' Roger North, 'a great Algibrist' and 'brother to the Lord North'.[61]

[59] See Dick (ed.), *Aubrey's Brief Lives*, pp. 24, 74, 97–100, 112, 115, 137–138, 188, 204, 229, 265, 353, 357, (possibly 369), 370–371, 400; Buchanan-Brown, *John Aubrey*, pp. 14–15, 47, 60, 78, 85, 123, 167, 190, 212–213, 239, 253, 267, 294, 322, 400; Hunter, *John Aubrey*, pp. 29, 31, 33 n.46, 98, 135, 198 n.8, 201; and Stephens (ed.), *Aubrey on Education*, pp. 85, 93, 101, 110, 195.

[60] See p. 202 n.5 *supra*.

[61] Dick (ed.), *Aubrey's Brief Lives*, p. 421, and Buchanan-Brown (ed.), *John Aubrey*, p. 221. FN's grandfather, the 3d lord North, was elder brother of the 'Captain'; see Randall, *Gentle Flame*, pp. 9–10 n.22, 11 n.1, 16–17 and n.48, 70 n.57

REFERENCES

Aarsleff, Hans
'John Wilkins (1614–72)', *Dictionary of Scientific Biography* ed. by C. C. Gillispie (16 vols., New York, 1970-80), vol. 14, pp. 361–381

Alexander, H. G. (ed.)
The Leibniz-Clarke Correspondence together with Extracts from Newton's Principia and Optics... (Manchester, 1956)

Alexander, Peter
Ideas, Qualities and Corpuscles: Locke and Boyle on the External World (Cambridge, 1985)

Anderson, Warren D.
Ethos and Education in Greek Music: The Evidence of Poetry and Philosophy (Cambridge, Massachusetts, 1966)

Andrews, H. K.
An Introduction to the Technique of Palestrina (London, 1958)

Arber, Edward
The Term Catalogues, 1668–1709 A.D., with a Number for Easter Term, 1711 A.D.: A contemporary Bibliography of English Literature in the Reigns of Charles II, James II, William and Mary, and Anne (3 vols., London, 1903)

Ariotti, P. E.
'Aspects of the Conception and Development of the Pendulum in the 17th Century', *Archive for History of Exact Sciences* (1971–72), 8: 329–410

Aristoxenus
The Harmonics ed. by H. S. Macran (Oxford, 1902)

Ashbee, Andrew (comp.)
Records of English Court Music (11 vols., Aldershot, 1986–96)

Ashbee, Andrew and David Lasocki
A Biographical Dictionary of English Court Musicians 1485–1714 (2 vols., Aldershot, 1998)

Atcherson, Walter
'Key and Mode in Seventeenth-Century Music Theory Books', *Journal of Music Theory* (1973), 17: 204–232

Backus, John
The Acoustical Foundations of Music (New York, 1969)

Baily, Francis
An Account of the Revd. John Flamsteed, the first Astronomer Royal compiled from his own Manuscripts, and other authentic Documents, never before published... (London, 1835)

Barbour, J. Murray
Tuning and Temperament: A Historical Survey (East Lansing, Michigan, 1953)

Barrow, Isaac
The Usefulness of Mathematical Learning explained and demonstrated: being Mathematical Lectures read in the publick Schools at the University of Cambridge ... tr. [from the 1st Latin edn. of 1685] by J. Kirkby (London, 1734)

Batten, M. I.
'The Architecture of Dr. Robert Hooke, F.R.S.', *Walpole Society* (1936–37), 25: 83–113 and plates

Békésy, Georg von
Sensory Inhibition (Princeton, 1967)

'Improved Musical Dynamics by Variation of Apparent Size of Sound Source', *Journal of Music Theory* (1970, published 1972), 4: 141–163

Beyer, Robert T.
Sounds of our Times: Two Hundred Years of Acoustics (New York, 1999)

Biggs, Hayes and Susan Orzel (eds.)
Musically Incorrect (New York, 1998)

Birch, Thomas
The History of the Royal Society of London ... in which the most Considerable of those Papers...which have hitherto not been published, are inserted [1756–57] (4 vols., London, 1968)

Birrell, T. A.
'Roger North and Political Morality in the later Stuart Period', *Scrutiny: A Quarterly Review* (1950–51), 27: 282–298

Boring, E. G.
Sensation and Perception in the History of Experimental Psychology (New York, 1942)

Boyle, Robert
The Works ... ed. by T. Birch (6 vols., London, 1965–66)

Brodsley, Laurel, Charles Frank and John W. Steeds
'Prince Rupert's Drops', *Notes and Records of the Royal Society* (1986), 41: 1–26

Brouncker, William
see Descartes

Brown, Joyce
'Guild Organisation and the Instrument-making Trade, 1550–1830: The Grocers' and Clockmakers' Companies', *Annals of Science* (1979), 36: 1–34

Bryden, D. J.
'Sir Samuel Morland's Account of the Balance Barometer, 1678', *Annals of Science* (1975), 32: 359–368

Buchanan-Brown, John (ed.)
John Aubrey: Brief Lives together with An Apparatus for the Lives of our English Mathematical Writers and The Life of Thomas Hobbes of Malmesbury (Harmondsworth, Middlesex, 2000)

Buchdahl, Gerd
'Gravity and Intelligibility: Newton to Kant', *The Methodological Heritage of Newton* ed. by R. E. Butts and J. W. Davis (Oxford, 1970), pp. 74–102

Burtt, E. A.
The Metaphysical Foundations of Modern Science (Garden City, New York, 1954)

Butler, Charles
The Principles of Musick, in Singing and Setting: with the two-fold Use therof, [ecclesiasticall and civil] (London, 1636)

Cannon, John T. and Sigalia Dostrovsky
The Evolution of Dynamics: Vibration Theory from 1687 to 1742 (New York, 1981)

Cassirer, Ernst
The Philosophy of Symbolic Forms (3 vols., New Haven, 1970)

Centore, F. F.
Robert Hooke's Contributions to Mechanics: A Study in Seventeenth Century Natural Philosophy (The Hague, 1970)

Chan, Mary and Jamie C. Kassler
'Francis North (1637–85)', *The New Grove Dictionary of Music and Musicians* ed. S. Sadie and J. Tyrrell (2d edn., 29 vols., London, 2000), vol. 18, pp. 52–53

Charleton, Walter
Physiologia Epicuro-Gassendo-Charltoniana: or a Fabrick of Science Natural, upon the Hypothesis of Atoms [1654] ... with indexes and a new introduction by R. H. Kargon (New York, 1966)

Chenette, Louis F.
Music Theory in the British Isles during the Enlightenment [doctoral dissertation, Ohio State University, 1967] (Ann Arbor, Michigan, 1968)

Chladni, E. F. F.
Die Akustik (Leipzig, 1802)

[Cluver, Dethlev]
'Claudii Ptolemæi *Harmonicorum Libri Tres. Ex Codd. Mss. Undecim, nunc primum Græce Editi*. Johannes Wallis...', *Philosophical Transactions of the Royal Society* (1683) [No. 143 January], 13: 20–21

Cohen, H. Floris
Quantifying Music: The Science of Music at the First Stage of the Scientific Revolution, 1580–1650 (Dordrecht, 1984)

Colie, Rosalie
Light and Enlightenment: A Study of the Cambridge Platonists and the Dutch Arminians (Cambridge, 1957)

Cromartie, Alan
Matthew Hale, 1609–1676: Law, Religion and Natural Philosophy (Cambridge, 1995)

Crombie, A. C.
'Marin Mersenne (1588–1648)', *Dictionary of Scientific Biography* ed. by C. C. Gillispie (16 vols., New York, 1970–80), vol. 9, pp. 316–322

Curwen, John
A Tract of Musical Static: An Attempt to show the Bearing of the most recent Discoveries in Acoustics on Chords, Discords, Transitions, Modulations, and Tuning, as used by modern Musicians (London, c.1867)

Dear, Peter
'Marin Mersenne: Mechanics, Music and Harmony', *Number to Sound: The Musical Way to the Scientific Revolution* ed. by P. Gozza (Dordrecht, 2000), pp. 267–288

Descartes, René
Discourse on Method, Optics, Geometry, and Meteorology [1637] tr. by P. J. Olscamp (Indianapolis, Indiana, 1965)

Excellent Compendium of Musick [tr. from the 1st Latin edn. of 1650]*: with Necessary and Judicious Animadversions Thereupon. By a Person of Honour* [William Brouncker] (London, 1653)

Treatise of Man tr. [from the 1662 Latin and 1664 French edns.] and commentary by T. S. Hall (Cambridge, Massachusetts, 1972)

Dethloff, Diana
'The Executors' Account Book and the Dispersal of Sir Peter Lely's Collection', *Journal of the History of Collections* (1996), 8: 15–51

Dick, O. L. (ed.)
Aubrey's Brief Lives (Harmondsworth, Middlesex, 1949)

Dickinson, H. W.
Sir Samuel Morland: Diplomat and Inventor, 1625–1695 (Cambridge, 1970)

Dijksterhuis, E. J.
The Mechanization of the World Picture tr. by C. Dikshoorn (London, 1969)

Dostrovsky, Sigalia
'Early Vibration Theory: Physics and Music in the Seventeenth Century', *Archive for the History of Exact Sciences* (1974–75), 14: 169–218

Dryden, John
Of Dramatick Poesie: An Essay ... and ... Defence of an Essay ... ed. by J. T. Boulton (Oxford, 1964)

Duverney, Joseph Guichard
Traité de l'organe de l'ouie, contenant la structure, les usages et les maladies de toutes les parties de l'oreille (Paris, 1683)

Eastwood, B. S.
'Descartes on Refraction: Scientific versus Rhetorical Method', *Isis* (1984), 75: 481–502

Enriques, Federigo
The Historic Development of Logic: The Principles and Structure of Science in the Conception of mathematical Thinkers tr. by J. Rosenthal (New York, 1929)

'Espinasse, Margaret
Robert Hooke (London, 1956)

Esterly, David
Grinling Gibbons and the Art of Carving (London, 1998)

Euler, Leonhard
Tentamen novæ theoriæ musicæ ex certissimis harmoniæ principiis... (St. Petersburg, 1739)

Evelyn, John
The Diary ... ed. by E. S. de Beer (6 vols., Oxford, 1955)

Eyre, G. E. B. (ed.)
A Transcript of the Registers of the Worshipful Company of Stationers, from 1640–1708 A.D. (3 vols., London, 1913–14)

Flamsteed, John
The Correspondence ... compiled and ed. by E. G. Forbes and (for Maria Forbes) by L. Murdin and F. Willmoth (3 vols., Bristol, 1995–2001)

Franklin, James
The Science of Conjecture: Evidence and Probability before Pascal (Baltimore, 2001)

Fussell, Paul
Theory of Prosody in Eighteenth-Century England (New London, Connecticut, 1954)

Galilei, Galileo
Dialogue concerning the Two Chief World Systems ... tr. [from the 1st Italian edn. of 1632] by S. Drake (Berkeley, 1953)

Dialogues concerning Two New Sciences tr. [from the 1st Italian edn. of 1638] by H. Crew and A. de Salvio... New York, 1954)

Gassendi, Pierre
Opera omnia ... ed. by H. L. Habert de Montmor and F. Henri (6 vols., Lyons, 1658)

Gaukroger, Stephen
Descartes: An Intellectual Biography (Oxford, 1995)

Gillispie, C. C.
The Edge of Objectivity: An Essay in the History of Scientific Ideas (Princeton, 1960)

Gouk, Penelope
'The Role of Acoustics and Music Theory in the Scientific Work of Robert Hooke', *Annals of Science* (1980), 37: 573–605

'Acoustics in the Early Royal Society 1660-1680', *Notes and Records of the Royal Society of London* (1982), 36: 155–175

Music, Science and Natural Magic in Seventeenth-Century England (New Haven, 1999)

Gozza, Paolo
'A Renaissance Mathematics: The Music of Descartes', *Number to Sound: The Musical Way to the Scientific Revolution* ed. by P. Gozza (Dordrecht, 2000), pp. 155–172

Grassby, Richard
The English Gentleman in Trade: The Life and Works of Sir Dudley North 1641–1691 (Oxford, 1994)

Grattan-Guinness, Ivor
'The Variety of Mechanics by 1800', *Historia Mathematica* (1990), 17: 313–338

Green, Burdette L.
The Harmonic Series from Mersenne to Rameau: An Historical Study of Circumstances leading to its Recognition and Application to Music [doctoral dissertation, Ohio State University, 1969] (Ann Arbor, Michigan, 1970)

Gruber, Albion
'Mersenne and Evolving Tonal Theory', *Journal of Music Theory* (1970), 14: 36–67

Hall, A. R.
'Mechanics and the Royal Society, 1668–70', *The British Journal for the History of Science* (1966), 3: 24–38

Hanson, N. R.
'Hypotheses fingo', *The Methodological Heritage of Newton* ed. by R. E. Butts and J. W. Davis (Oxford, 1970)

Harding, R. E. M.
Origins of Musical Time and Expression (London, 1938)

Helmholtz, Hermann von
On the Sensations of Tone as a Physiological Basis for the Theory of Music ... tr. [from the 4th German edn. of 1877] by A. J. Ellis [1885]... (New York, 1954)

Herissone, Rebecca
Music Theory in Seventeenth Century England (Oxford, 2000)

Herschel, J. F. W.
'Sound' [1830], *Encyclopædia metropolitana...* (27 vols., London, 1817–45), vol. 4, pp. 747–825 and plates, vol. 1

Hesse, Mary
Forces and Fields: The Concept of Action at a Distance in the History of Physics (Westport, Connecticut, 1970)

'Hooke's Vibration Theory and the Isochrony of Springs', *Isis* (1966), 57: 533–441

Higgins, W. M.
The Philosophy of Sound and Musical Composition (London, 1838)

Hobbes, Thomas
'Human Nature' [1650], *The English Works* ... ed. by W. Molesworth (11 vols., London, 1839–45), vol. 4, pp. 1–76

'Elements of Philosophy. The first Section, concerning Body' [1656], *The English Works* ... ed. by T. Molesworth (11 vols., London 1839–45), vol. 1

Holder, William
'An Account of an Experiment, concerning Deafness...', *Philosophical Transactions of the Royal Society* (1668) [No. 35, 18 May], 3: 665–668

A Treatise of the Natural Grounds, and Principles of Harmony (London, 1694)

Holman, Peter
'Suites by Jenkins rediscovered', *Early Music* (1978), 6: 25–55

Hooke, Robert
Micrographia or Some Physiological Descriptions of Minute Bodies made by Magnifying Glasses with Observations and Inquiries thereupon [1665] (New York, 1961)

'A Description of Helioscopes, and Some Other Instruments...' [1676], R. T. Gunther, *Early Science in Oxford* (15 vols., Oxford, 1923–67), vol. 8, pp. 119–208

Lectures de potentia restitutiva, or of Spring explaining the Power of Springing Bodies...' [1678], R. T. Gunther, *Early Science in Oxford* (15 vols., Oxford, 1923–67), vol. 8, pp. 331–388

The Diary [1672–80] ... ed. by H. W. Robinson and W. Adams... (London, 1935)

'The Diary' [1688–93], transcribed by R. T. Gunther, *Early Science in Oxford* (15 vols., Oxford, 1923–67), vol. 10, pp. 69–265

The Posthumous Works ... containing his Cutlerian Lectures and other Discourses, read at the Meetings of the illustrious Royal Society [ed. by R. Waller 1705]... (London, 1971)

'The Life and Work of Robert Hooke', transcribed by R. T. Gunther, *Early Science in Oxford* (15 vols., Oxford, 1923–67), vols. 6 and 7

Hoppen, K. T.
The Common Scientist in the Seventeenth Century: A Study of the Dublin Philosophical Society 1683–1708 (Charlottesville, Virginia, 1970)

Hunt, Frederick Vinton
Origins in Acoustics: The Science of Sound from Antiquity to the Age of Newton (New Haven, 1978)

Huygens, Christian
Œuvres complètes... (22 vols., The Hague, 1888–1950)

Treatise on Light. In which are explained the Causes of that which occurs in Reflexion, & in Refraction [1690] ... tr. by S. P. Thompson (New York, 1962)

Hyer, Brian
'Tonality', *New Grove Dictionary of Music and Musicians* ed. by S. Sadie and J. Tyrrell (2d edn., 29 vols, London, 2000), vol. 25, pp. 583–594

Jacob, C. T.
'Notes on Newton's Letter of 1677 to the Hon. and Rev. John North', *Hinrichsen's Musical Year Book* (1947–48), 4–5: 479–481

Jones, Howard
The Epicurean Tradition (London, 1992)

Josten, C. H. (ed.)
Elias Ashmole (1617–1692): his autobiographical and historical Notes, his Correspondence, and other contemporary Sources relating to his Life and Work (5 vols., Oxford, 1966)

Kalmar, Martin
'Thomas Hariot's *De Reflexione Corporum Rotundorum*: An Early Solution to the Problem of Impact', *Archive for History of Exact Sciences* (1977), 16: 201–230

Kassler, Jamie C.
The Science of Music in Britain, 1714–1830: A Catalogue of Writings, Inventions and Lectures (2 vols., New York, 1979)

Inner Music: Hobbes, Hooke and North on Internal Character (London, 1995)

'Joshua Steele (1700–91)', *New Grove Dictionary of Music and Musicians* ed. by S. Sadie and J. Tyrrell (2d edn., 29 vols., London, 2000), vol. 24, pp. 309–310

'Roger North (1651–1734)', *New Grove Dictionary of Music and Musicians* ed. by S. Sadie and J. Tyrrell (2d edn., 29 vols., London, 2000), vol. 18, pp. 53–56

Music, Science, Philosophy: Models in the Universe of Thought (Aldershot, 2001)

Kassler, Jamie C. and David R. Oldroyd
'Robert Hooke's Trinity College "Musick Scripts", his Music Theory and the Role of Music in his Cosmology', *Annals of Science* (1983), 40: 559–595

Keele K. D.
Leonardo da Vinci's Elements of the Science of Man (Academic Press, 1983)

Kircher, Athanasius
Musurgia universalis sive ars magna consoni et dissoni in X libros digesta (2 vols., Rome, 1650)

Kitchin, George
Sir Roger L'Estrange: A Contribution to the History of the Press in the Seventeenth Century (London, 1913)

Korsten, F. J. M.
Roger North (1651–1734): Virtuoso and Essayist (Amsterdam, 1981)

Kraemer, Alan
'Two Speakers are better than 5.1', *IEEE Spectrum* (May 2001), pp. 71–74

Lenihan, J. M. A.
'Mersenne and Gassendi', *Acustica* (1951), 1: 96–99

Lenoble, Robert
Mersenne ou la naissance du mécanism (Paris, 1943)

Letwin, William
The Origins of Scientific Economics: English Economic Thought 1660–1776 (London, 1963)

Lindsay, R. B.
'The Story of Acoustics', *The Journal of the Acoustical Society of America* (1966), 39: 629–644

Lysaght, D. J.
'Hooke's Theory of Combustion', *Ambix* (1937), 1: 93–108

McCrory, Edmund
A few Notes on the Temple Organ (2d edn., London, c.1861)

Macdonald, Hugh and Mary Hargreaves
Thomas Hobbes: A Bibliography (London, 1952)

Mace, Thomas
Musick's Monument; or, a Remembrancer of the best practical Musick, both divine, and civil, that has ever been known, to have been in the World... (London, 1676)

McKie, Douglas
'Some early Work on Combustion, Respiration and Calcination', *Ambix* (1938), 1: 143–165

Maley, V. Carlton, Jr.
The Theory of Beats and Combination Tones 1700–1863 (New York, 1990)

Manley, Lawrence
Convention 1500–1750 (Cambridge, Massachusetts, 1980)

Mariotte, Edme
Traité de la percussion ou choq des corps, dans lequel les principales regles du mouvement contraines a celles que M. Descartes, & quelques autres modernes ont voulu establir, sont demonstrées par leur véritables causes (Paris, 1673)

Matthews, M. R.
Time for Science Education: How Teaching the History and Philosophy of Pendulum Motion can contribute to Science Literacy (New York, 2000)

Mersenne, Marin
Harmonie universelle: The Books on Instruments tr. [from the French edn. of 1636] by R. E. Chapman (The Hague, 1957)

Messen, John
'A Question of Temperament: Purcell and Croft', *The Musical Times* (1978), 119: 504–506

Middleton, W. E. Knowles
The Experimenters: A Study of the Accademia del Cimento (Baltimore, 1971)

The History of the Barometer (Baltimore, 1964)

Millar, Oliver
Sir Peter Lely 1618–80: Exhibition at 15 Carlton House Terrace... (London, 1978)

Miller, George A.
'The Magical Number Seven, Plus or Minus Two: Some Limits on Our Capacity for Processing Information', *The Psychology of Communication: Seven Essays* (London, 1968), pp. 14–44

Morley, Thomas
A Plaine and Easie Introduction to Musick, set downe in forme of a Dialogue devided into three Parts... (London, 1597)

Moss, Ann
Printed Commonplace-Books and the Structuring of Renaissance Thought (Oxford, 1996)

Mullett, C. F.
'"An Arte to Make the Dumbe to Speake, the Deafe to Heare": A Seventeenth-Century Goal', *Journal of the History of Medicine and Allied Sciences* (1971), 26: 123–149

Newton, Isaac
Mathematical Principles of Natural Philosophy ... tr. [from the Latin] by A. Motte [1729] and revised by F. Cajori (Berkeley, 1946)

Opticks or A Treatise of the Reflections, Refractions, Inflections & Colours of Light [based on the 4th edn. of 1730]... (New York, 1952)

Papers & Letters on Natural Philosophy and related Documents ed. by I. B. Cohen and R. E. Schofield (Cambridge, Massachusetts, 1958)

The Correspondence... (7 vols., Cambridge, 1959–77)

[North, Dudley]
Discourses upon Trade; principally directed to the Cases of the Interest, Coynage, Clipping, Increase of Money (London, 1691)

[North, Francis]
'A Conjecture concerning the Bladders of Air that are found in Fishes, communicated by A.I. [=Anonymous Individual]; and illustrated by an Experiment suggested by the Honorable Robert Boyle', *Philosophical Transactions of the Royal Society* (1675) [No. 114, 24 May], 10: 310–311

North, Marianne
Recollections of a Happy Life being the Autobiography ed. by her sister Mrs John Addington Symonds (2 vols., London, 1892)

North, Roger
A Discourse on the Study of the Laws ... with notes and illustrations by a Member of the Inner Temple (London, 1824)

'Letters of the Hon[oura]ble Roger North', *The Autobiography* ... ed. A. Jessopp (London, 1887), appendix ii, pp. 221–281

Life/DN = 'The Life of the Hon. Dudley North', *The Lives of the Norths* ed. by A. Jessopp (3 vols., London 1890), vol. 2

Of Building: Roger North's Writings on Architecture ed. by H. Colvin and J. Newman (Oxford, 1981)

Life/JN = *General Preface & Life of Dr John North* ed. by P. Millard (Toronto, 1984)

Cursory Notes of Musicke (c.1698–c.1703) ... ed. by M. Chan and J. C. Kassler (Kensington, NSW, 1986)

The Musicall Grammarian 1728 ed. by M. Chan and J. C. Kassler (Cambridge, 1990)

Life/FN = The Life of the Lord Keeper North ed. by M. Chan (Lewiston, 1995)

Notes of Me: The Autobiography of Roger North ed. by P. Millard (Toronto, 2000)

North Papers
NP 2 = Roger North's Writings on Music to c.1703: A Set of Analytical Indexes prepared by J. D. Hine *with Digests of the Manuscripts* by M. Chan and J. C. Kassler (Kensington, NSW, 1986)

NP 3 = Roger North's The Musicall Grammarian and Theory of Sounds: Digests of the Manuscripts by M. Chan and J. C. Kassler *with an Analytical Index of 1726 and 1728 Theory of Sounds* by J. D. Hine (Kensington, NSW, 1988)

NP 4 = Roger North: Materials for a Chronology of his Writings [Checklist No. 1] by M. Chan and J. C. Kassler (Kensington, NSW, 1989)

NP 5 = Roger North's Writings on Music c.1704–c.1709: Digests of the Manuscripts by M. Chan and J. C. Kassler *with Analytical Indexes* by J. D. Hine (Kensington, NSW, 1999)

NP 6 = Roger North's 'Of Sounds' and Prendcourt Tracts c.1710–c.1716: Digests and Editions by M. Chan and J. C. Kassler *with an Analytical Index* by J. D. Hine (Kensington, NSW, 2000)

Oldenburg, Henry
The Correspondence ... ed. and tr. by A. R. Hall and M. B. Hall (13 vols., London, 1965–86)

Omond, T. S.
A Study of English Metre (London, 1903)

English Metricists in the Eighteenth and Nineteenth Centuries being a Sketch of English prosodical Criticism during the last two hundred Years (London, 1907)

Parker, R. H.
'Roger North: Gentleman, Accountant and Lexicographer', *Accounting History* (1997), 2: 31–51

Pepys, Samuel
Private Correspondence and Miscellaneous Papers ed. by J. R. Tanner (2 vols., London, 1926)

The Diary ... ed. by R. Latham and W. Matthews (11 vols., Berkeley, 1970)

Pinkerton, R. C.
'Information Theory and Melody', *Scientific American* (1956), 194: 77–86

Plomer, H. R.
A Dictionary of Booksellers and Printers who were at Work in England, Scotland and Ireland from 1641 to 1667 (Oxford, 1968)

Poole, H. Edmund
'The Printing of William Holder's "Principles of Harmony"', *Proceedings of the Royal Musical Association* (1974–75), 101: 31–43

Purver, Margery
The Royal Society: Concept and Creation (Cambridge, Massachusetts, 1967)

Randall, D. B. J.
Gentle Flame: The Life and Verse of Dudley, Fourth Lord North (1602–1677) (Durham, North Carolina, 1983)

Ray, John
'A Letter written to the Publisher ... containing some Considerations on the Conjecture in Numb. 114 of these Tracts, about the swimming Bladders in Fishes', *Philosophical Transactions of the Royal Society* (1675) [No. 115, 21 June], 10: 349–351

Reddaway, T. F.
The Rebuilding of London after the Great Fire (London, 1940)

Reilly, Conor
'Francis Line, Peripatetic', *Osiris* (1962), 14: 222–253

Roberts, Francis
'A Discourse concerning the Musical Notes of the Trumpet, and Trumpet-Marine, and of the Defects of the Same', *Philosophical Transactions of the Royal Society* (1686–92) [No. 195, 19 October 1692], 16: 559–563

Rochot, Bernard
'Pierre Gassendi (1592–1655)', *Dictionary of Scientific Biography* ed. by C. C. Gillispie (16 vols., New York, 1970–80), vol. 5, pp. 284–290

Roederer, Juan G.
Introduction to the Physics and Psychophysics of Music (2d edn., New York, 1975)

Rostenberg, Leona
The Library of Robert Hooke: The Scientific Book Trade of Restoration England (Santa Monica, California, 1989)

Sampson, Margaret
'Laxity and Liberty in seventeenth-century English Political Thought', *Conscience and Casuistry in Early Modern Europe* ed. by E. Leites (Cambridge, 1988), pp. 72–118

Schneewind, J. B.
The Invention of Autonomy: A History of Modern Moral Philosophy (Cambridge, 1998)

Scott, W. L.
The Conflict between Atomism and Conservation Theory 1644 to 1860 (London, 1970)

Searle, Humphrey
Twentieth Century Counterpoint... (London, 1954)

Shannon, Claude E. and William Weaver
The Mathematical Theory of Communication (Urbana, Illinois, 1949)

Shapiro, Alan E.
'Kinematic Optics: A Study of the Wave Theory of Light in the Seventeenth Century', *Archive for History of Exact Sciences* (1974), 11: 134–266

'Light, Pressure, and Rectilinear Propagation: Descartes' Celestial Optics and Newton's Hydrostatics', *Studies in History and Philosophy of Science* (1974), 5: 239–296

'Artists' Colors and Newton's Colors', *Isis* (1994), 85: 600–630

'The Gradual Acceptance of Newton's Theory of Light and Color, 1672–1727', *Perspectives on Science* (1996), 4: 59–140

Shapiro, Barbara
Probability and Certainty in Seventeenth-Century England: A Study of the Relationships between Natural Science, Religion, History, Law, and Literature (Princeton, 1983)

Shirlaw, Matthew
The Theory of Harmony: An Inquiry into the Natural Principles of Harmony, with an Examination of the Chief Systems of Harmony from Rameau to the Present Day (DeKalb, Illinois, 1955)

Simpson, Christopher
The Division-Viol or the Art of Playing extempore upon a Ground [2d edn., 1665] (London, 1965)

A Compendium of Practical Musick [1667] *in five Parts...* (3d edn., London, 1678)

Smith, Robert
Harmonics, or The Philosophy of Musical Sound (London, 1749)

Sotheby's
Printed and Manuscript Music (Sale L00202, London, 26 May 2000)

Spink, Ian
'The Old Jewry "Musick-Society": A 17th-Century Catch Club', *Musicology* [Journal of the Musicological Society of Australia] (1965–67), 2: 35–41

Stephens, J. E. (ed.)
Aubrey on Education: A hitherto unpublished Manuscript by the Author of Brief Lives (London, 1972)

Taylor, C. A.
The Physics of Musical Sounds (London, 1965)

Thoren, V. E.
'John Flamsteed (1646–1719)', *Dictionary of Scientific Biography* ed. by C. C. Gillispie (16 vols., New York, 1970–80), vol. 5, pp. 22–26

Tilmouth, Michael
'A Calendar of References to Music in Newspapers published in London and the Provinces (1660–1719)', *R.M.A. Research Chronicle* (1961), 1: vii, 1–107, 2: 1–15

Timoshenko, Stephen P.
History of Strength of Materials: With a brief Account of the History of Theory of Elasticity and Theory of Structures (New York, 1953)

Tonndorf, Jurgen (ed.)
Physiological Acoustics (Stroudsburg, Pennsylvania, 1981)

Truesdell, Clifford
The Rational Mechanics of Flexible or Elastic Bodies 1638–1788... (Turin, 1960)

Voss, Stephen
'Scientific and Practical Certainty in Descartes', *American Catholic Philosophical Quarterly* (1993), 67: 569–584

Walker, D. P.
Studies in Musical Science in the Late Renaissance (London, 1978)

Wallis, John
'An Extract of Letters ... to the Publisher, 1672. Sept. 26. &c. concerning the Suspension of Quicksilver well purged of Air, much higher than the ordinary Standard in the Torricellian Experiment', *Philosophical Transactions of the Royal Society* (1672–73) [No. 91, 24 February 1673], 7: 5160–5170

'Dr. Wallis's Letter to the Publisher, concerning a new musical Discovery; written from Oxford, March 14. 1676/7', *Philosophical Transactions of the Royal Society* (1677) [No. 134, 23 April], 12: 839–842

'Appendix de veterum harmonica ad hodiernam comparata', *Claudii Ptolemæi Harmonicorum libri tres* (Oxford, 1682), pp. 281–328

Opera mathematica... (3 vols., Oxford, 1693–99)

Webster, Charles
'The Discovery of Boyle's Law, and the Concept of the Elasticity of Air in the Seventeenth Century', *Archive for the History of Exact Sciences* (1962–66), 2: 441–502

Wedgwood, J. I.
A Comprehensive Dictionary of Organ Stops English and Foreign, Ancient and Modern... (London, c.1905)

Westfall, Richard
Force in Newton's Physics: The Science of Dynamics in the Seventeenth Century (London, 1971)

Whiteside, D. T.
'Isaac Barrow (1630–77)', *Dictionary of Scientific Biography* ed. by C. C. Gillispie (16 vols., New York, 1970–80), vol. 1, p. 475

Wiener, Norbert
Cybernetics or Control and Communication in Animal and Machine (2d edn., New York, 1961)

Wienpahl, R. W.
'Modality, Monality and Tonality in the Sixteenth and Seventeenth Centuries', *Music & Letters* (1971), 52: 407–417, (1972), 53: 59–73

Willets, P. J.
'Autograph Music by John Jenkins', *Music & Letters* (1967), 43: 124-126

Williams, E.
'Hooke's Law and the Concept of the Elastic Limit', *Annals of Science* (1956), 12: 74–83

Willmoth, Frances
Sir Jonas Moore: Practical Mathematics and Restoration Science (Woodbridge, Suffolk, 1993)

Wilson, John
Roger North on Music: Being a Selection from his Essays written during the Years c.1695–1728 (London, 1959)

Winckel, Fritz
Music, Sound and Sensation: A Modern Exposition tr. [from the revised German edn. of 1960] by T. Binkley (New York, 1967)

Wood, Alexander
The Physical Basis of Music (Cambridge, 1913)

Young, Matthew
An Enquiry into the Principal Phænomena of Sounds and Musical Strings (Dublin, 1784)

INDEX OF NAMES

Academicians *see* Accademia del Cimento
Accademia del Cimento 19, 44–5
Adamson, Humphry (*fl.* 1670s) 208–209
Anonymous critic vii, 91–2, 95, 131, **171–74**
Aristotle (384–322 B.C.) 7
Aristotelian author 55
Aristoxenus (*fl.* c.318 B.C.) 78
Arminius, Jacobus (1560–1609) 34
Ashmole, Elias (1617–92) 201
Atcherson, Walter 81
Atomists 5
see also Epicureans
Aubrey, John (1626–97) 23, 36, 135, 199, 201–203, 205, 208, 210–11

Babbitt, Milton 129
Ball, Peter (*fl.* 1660s) 202
Ball, Peter (d. 1675) 202
Ball, William (d. 1690) 202–203
Banks, Joseph (1743–1820) 131
Barbour, J. Murray 81
Barrow, Isaac (1630–77) 2, 17, 30, 34–5, 89, 121
Beale, Simon (1660–80) 209
Beeckman, Isaac (1588–1637) 5
Blagge, Margaret *see* Godolphin, Margaret
Boyle, Robert (1627–91) 15–16, 18, 19–20, 26–7, 42–4, 60, 86, 110, 121, 175, 209
Brouncker, William (1620–84) 2d Viscount 18, 30–31, 80, 129, 174
Bulwer, John (*fl.* 1600–50) 58
Butler, Charles (d. 1647) 77
Butler, Samuel (1612–80) 29

Cambridge Platonists 34
see also Platonists
Cassirer, Ernst 123, 130
Cavendish, Margaret (d. 1673) duchess of Newcastle 25
Charleton, Walter (1619–1707) 5–8, 10, 15, 19, 30, 42, 44–5, 49, 58, 98
Charles II (1630–85) xiii, 23, 29, 30, 31, 34, 61, 205
as duke of York 207
Cluer, Dethlev (c.1650–1708) 89
Cohen, H. Floris 52
Coke, Edward (1552–1634) 39
Colvin, Howard 109
Compton, Henry (1632–1713) 29
Cooper, Samuel (1609–72) 205
Copernicus, Nicolaus (1473–1543) 53
Corporation for Regulating the Art and Science of Music 23
Cotes, Roger (1682–1716) 103–104
Cotton, Robert (d. 1712) 207

Dale, Henry H. 134
Derham, William (1657–1735) 110
Descartes, René (1596–1650) 8–10, 13, 15, 17–20, 31, 34, 37, 46, 51–2, 55, 79–80, 84, 94, 107, 115
Dostrovsky, Sigalia 74
Duport, James (1606–79) 34
Duverney, Joseph Guichard (1648–1730) 124

Edgcumbe, Richard (1640–88) 35
Ellis, Alexander J. 1, 79, 81, 99, 125, 148
Epicureans 5, 10
Euler, Leonhard (1707–83) 56
Evelyn, John (1620–1706) 29–30, 202–203

Finch, Edward (1664–1738) 26
Finch, Heneage (1621–82) 1st earl of Nottingham 26
Finch, John (1626–82) 26
Flamsteed, John (1646–1719) 203–204, 206–207
Fourier, J. B. J. (1768–1830) 126

Gale, Thomas (1635?–1702) 33–4, 37

Galilei, Galileo (1564–1642) 3, 10, 13–14, 46, 49, 52, 58, 63–4, 71, 74, 94, 98, 122–24
Gassendi, Pierre (1592–1655) 5, 7, 19–20, 44, 49
Gayer, Mary *see* North, Mary
Gayer, Robert (d. 1702) xiii, 207
Gentleman's Society 101
Gibbons, Grinling (1648–1720) 206–207
Godolphin, Margaret *née* Blagge (1652–78) 29
Godolphin, Sidney (1645–1712) 1st earl of Godolphin 29
Green, Burdette 62, 73
Guericke, Otto von (1602–86) 176

Hale, Matthew (1609–76) 20, 26–7, 38, 108, 209
Harris, Renatus (c.1652–1724) 61, 115
Hartlib, Samuel (d. 1670?) 210
Hauksbee, Francis (d. 1713) 95
Helmholtz, Hermann von (1821–94) 1–2, 4, 51, 55, 70, 75, 82, 113, 118, 121–22, 124, 126, 128–30
Hesse, Mary 20
Hill, Abraham (1635–1721) 36, 71
Hindemith, Paul 129
Hobbes, Thomas (1588–1679) vii, 2, 3, 10–18, 20, 31, 40, 45, 46, 48, 52, 53, 58–9, 76–8, 85, 88, 92, 100, 107, 111, 118, 122, 126, 186, 202–203, 205–206, 208
Holder, William (1616–98) 36, 58, 70, 89–91
Hooke, Robert (1635–1702) 2, 18, 20–21, 27–8, 30, 33, 35–6, 41, 46–50, 53, 56–7, 60–61, 70–75, 86, 89–90, 92–4, 97–8, 101, 105–107, 111–13, 115, 117–18, 122–25, 131, 202–203, 205–206, 208–11
Hoskins, John (1634–1705) 35–6, 202, 204–205
Huygens, Christian (1629–95) 16–18, 20, 58, 84, 95, 106–7, 110–11, 203, 205, 208
Hunt, Frederick Vinton 74, 95–6, 121

Inner Temple *see* Temple

Jacob, C. T. 95–100
James II (1635–1701) xiii, 23, 127
as duke of York 29, 208, 210
Jenkins, John (1592–1678) 23
Jones, Henry (*fl.* 1654–92) 209
Jones, Richard (1636–1712) 3d Viscount and 1st earl of Ranelagh 26

Kargon, Robert 44
Kepler, Johannes (1571–1630) 123
King, John (1639–77) 29

Leibniz, Gottfried Wilhelm (1646–1716) 125
Lely, Peter (1618–80) xii–xiv, 205–207
L'Estrange, Roger (1616–1704) 29
Lilly, John (d. 1678) 23
Line, Francis (1595–1675) 27, 209
Littleton, Thomas (1402–81) 39
Loggan, David (d. 1692) x
Longueville, William (1639–1721) 29
Loosemore, George (d. 1678) 23, 61
Loosemore, Henry (d. 1670) 23, 61
Loosemore, John (d. 1681) 61
Lully, Jean-Baptiste (1632–87) 101

Mace, Thomas (d. 1706?) 58
Maley, Carlton 56, 78
Marsh, Narcissus (1638–1713) 69
Martyn, John (c.1619–80) 25, 28, 171
Martyn, Sarah (*fl.* 1683) 25
Mary of Modena (1658–1718) xiii
Matteis, Nicola (*fl.* 1670–1714) 25
Matthews, Edmund (d. 1692) 23, 89
May, Baptist (1629–98) 206
May, Hugh (1622–84) 205–207
Mersenne, Marin (1588–1648) 2, 44–5, 49, 52, 55, 58, 67–8, 74, 94, 121, 183
Middle Temple *see* Temple
Millard, Peter T. 23, 27, 29, 30, 32, 108, 206
Montagu, Anne *see* North, Anne
Montagu, Charles (d. 1625) [*grandfather of* FN] xi
Montagu, Edward (1532–1602) 1st baron Montagu of Boughton [*great grandfather of* FN] xi
Montagu, Oliver (d. 1693) 116
Montagu, Ralph (1638?–1709) 1st duke of Montagu 35

Moore, Jonas (1617–79) xii, 202–204, 106–107, 209–10
Morland, Samuel (1625–95) 89, 206, 208–209
Morley, Thomas (c.1557–1602) 163
Musick Society 23

New Philosophical Club 202, 205, 208, 211
Newman, John 109
Newton, Isaac (1642–1727) vii, 16, 18, 28, 32, 35, 38, 60, 89, 91–2, 94–100, 102–106, 112–13, 121–22, 127–28, 131–34, **175–78**
Noble, William (*fl.* 1670–73) 69
North, Anne *née* Montagu (c.1613–81) [*mother of* FN] xi, 33
North, Anne (d. 1726) [*daughter of* FN] xii
North, Charles (1635–91) [*eldest brother of* FN] 23
North, Charles (1676–1714) [*son of* FN] xii
North, Dudley (1581–1666) 3d baron North [*grandfather of* FN] xii–xiii, 23, 61, 211
North, Dudley (1602–77) **DN (1)** 4th baron North [*father of* FN] xi–xiii, 23, 28, 32–3, 199, 202, 205, 207
North, Dudley (1641–91) **DN (2)** [*brother of* FN] xii–xiii, 27, 73, 88, 104–5, 112, 206, 210
North family xi, 33, 35, 61, 131, 203
North, Fountain (1749–c.1810) 134
North, Frances *née* Pope (1647–78) [*wife of* FN] xii, 76, 206
North, Francis (1637–85) [**FN**] 1st baron Guilford
  biographical xii–xiii, 21–39, 44, 51, 61, 73, 80, 87, 89, 101, 104, 109, 116, 135, 178, 201–11
  miscellaneous writings 26–7, 31–2, 38–9, 47, 61, 140, 183, 208–209
  A Philosophical Essay of Musick vii, 4, 21, 25, 28, 30, 32, 33, 35, 37–87, 89–117, 121, 123–24, 126–29, 133, **135–70**, 171, 175, 179, 201, 209
North, Francis (1673–1729) [*son of* FN] xii, 61
North, Frederick (1800–69) 134
North, Frederick Francis (1778–1821) 134
North, Janet *see* Symonds, Janet North
North, John (1645–83) **JN** [*brother of* FN] xii–xiii, 28–30, 32–7, 94, 104, 134, 175, 177
North, Marianne (1830–90) 134
North, Mary *née* Gayer (d. 1729?) xiii
North, Montagu (1712–79) xiv, 21
North, Robert (*fl.* 1660–80) 30
North, Roger (*fl.* 1619–51) 211
North, Roger (1651–1734) **RN** [*youngest brother of* FN] vii, ix, xi–xiv, 25–7 31, 34–5, 37–40, 61, 89, 91, 121, 123, 126–27, 201, 207–209
  biographical x–xii, 21, 25, 27, 30, 32, 36, 39, 56, 73, 84, 87, 101, 103–109, 112, 115–17, 134, 206–207
  develops hints in FN's *Essay* 56, 76–7, 80, 82–4, 88, 112–15, 153, 163
  miscellaneous writings 21, 27, 32, 39, 44, 49, 56, 72–7, 81, 83–4, 88, 102–103, 105, 107–13, 116–17, 119–20, 127, 148, 153, 156, 163, 182–83, 185–88, 193, 208–209
  'Some Notes upon an Essay' 21, 91, 104–105, 108–20, 126–27, 131, 133, **179–200**
North, Roger (1703–71) 134
North, Roger (1901–86) 134
North, Thomas Frederick 134

Ohm G. S. (1789–1854) 126
Oldenburg, Henry (d. 1677) 26–7, 97, 209
Osborne, Thomas (1631–1712) 1st earl of Danby 29

Palmer, Edward (*fl.* 1660s) 30
Palmer, Geoffrey (1598–1670) 29–30
Parker, Samuel (1640–88) 29
Pawlett, Robert (d. 1683) 28–9
Pearson, John (1613–86) 34
Pepys, Samuel (1633–1703) 66, 69, 89, 148, 208
Petty, William (1623–87) 210–11
Pigot, George (*fl.* 1659–72) 23
Pigot, Thomas (d. 1686) 69
Plato 37–8

Platonists 38, 119
see also Cambridge Platonists
Playford, John (1623–86) 23, 209
Pope, Frances *see* North, Frances
Pope, Thomas (1598–1668) 3d earl of Downe xii
Prin, Jean-Baptiste (*fl.* 1674–1742) 66
Prin, Monsieur (*fl.* 1667–74) 66
Ptolemaeus, Claudius (*fl.* 139–161) 89
Ptolemy *see* Ptolemaeus, Claudius
Purcell, Henry (1659–95) 25
Pythagoreans 52, 104

Riley, John (1646–91) xii, 22
Roberts, Francis (1650–1718) 101–102
Roberts, John (1606–85) earl of Radnor 101
Rooke, Lawrence (1622–62) 34
Royal Society vii, 16, 25–8, 30–31, 33–5, 41, 56, 69, 71, 90–91, 101, 124, 134, 171, 201–205, 207, 209–10

Sabine, Edward (1788–1883) 134
Sabine, Wallace Clement 150
Sauveur, Joseph (1653–1716) 53, 55, 100, 121
Savart, Félix (1791–1841) 74
Sawyer, Robert (1633–92) 116
Scholastics 7, 10
Shannon, Claude E. 119
Shapiro, Alan E. 17, 97
Shore family 153
Shore, John (c.1662–1752) 153
Sill, William (d. 1687) 29
Simpson, Christopher (d. 1669) 80–81, 163
Smith, Bernard (c.1630–1708) 61, 81 115–17
Southwell, Robert (1635–1702) 26
Stokeham, William (d. 1698) 207
Strutt, J. W. (1842–1919) lord Rayleigh 99
Symonds, Janet North 134
Symonds, John Addington, Jr. (1849–93) 134

Temple 29–30, 38, 210
Church 23, 81
Inner Temple 29, 115
Middle Temple xiii–xiv, 21, 29–30, 115, 201–203, 205, 208–10
Tilson, Henry (1659–95) xiii
Tompion, Thomas (1639–1713) 71
Torricelli, Evangelista (1608–47) 19, 110
Townley, Richard (1629–1707) 60–61
Truesdell, Clifford 73
Turnbull, H. W. 25, 134

Vaux, Theodore de (d. 1694) 205
Villiers, George (1628–87) 2d duke of Buckingham 206
Vitruvius Pollio (c.84–14 B.C.) 153, 173
Vossius, Isaac (1618–89) 183

Waller, Richard (d. 1714) 71–2, 74
Wallis, John (1616–1703) 16–18, 69–70, 89–91, 101–102, 183
Werden, John (1640–1716) 207, 209–10
Werden, Robert (d. 1690) 209
Wilkins, John (1614–72) 2
Williams, E. 92
Williamson, Joseph (1623–1701) 26
Willis, Thomas (1621–75) 76
Wilson, John 31–2
Wren, Christopher (1632–1723) 16, 28, 34, 36, 70–71, 73, 92, 112
Wyld, Edmund (*fl.* 1649–93) 202, 207, 210–11
Wynne, Henry (d. 1709) 209

York, duke of *see* Charles II, James II
Young, Matthew (1750–1800) 103

INDEX OF SUBJECTS

Accelerating force *see* Force
Accent *see* Rhythm
Account 40, 45, 50, 54, 74–5, 77, 85, 87, 117–18, 126, 162, 184, 190
  accounts 88n.
  calculate 187n.
  unaccountable 186
Acoustic levers *see* Resonators
Acoustic signals 1, 3, 117
Acoustic spectrum 56, 60, 116
Acoustic vibrators *see* Vibrators
Acoustics 2–4, 10n., 17, 59, 94, 105, 107, 121–22, 124, 150n.
  see also Physics
Acquiescence *see* Equilibrium
Aerophones 58–61, 68–9, 92, 100, 109, 113
  wind instruments 51, 58, 148n., 196
  see also Flutes and flue pipes; Musical instruments; Organs; Pipes
Affect 3
  feelings 1–2, 77–8, 85, 129, 184n.
  passions 3, 14–15, 21, 85, 163
  see also Experiences; Pain; Pleasure
Air 30, 32, 41
  as mixture 18, 45–7, 139, 171
  constituent parts 5, 15, 17–18, 27, 47
  figure of 3, 17
  physical properties 18–20, 45–7, 110–11, 139–40, 171, 179–80
  see also Material medium
Air columns 56, 62, 68, 71, 94, 115, 123, 148n., 149n., 154, 181n., 199
  see also Aerophones; Flutes and flue pipes; Pipes
Air pumps 42, 47n., 121, 175n.
Air or water resistance 46, 57, 96
Amplification *see* Resonance; Sound reinforcement and destruction
Amplitude *see* Vibration
Analogues 37, 48, 65, 198
  ball 17, 37
  water wave 12, 17, 97–8, 106, 109, 145, 177, 179, 193
  see also Models; Pendula
Anticipation *see* Conception
Architecture and building 109, 186, 187n., 188–90
Arminianism 34
Atmosphere *see* Air
Attention *see* Mental faculties
Attention process 1, 77–8, 82–6, 119, 124, 126, 155–56, 161–64, 185
  see also Conception; Information processing
Audibility range 21, 49–50, 117, 124, 141, 191–92
  see also Hearing
Auditory localisation 99, 102–3, 121, 177–78
Auditory nerves *see* Brain and nervous system
Auditory system 9, 11–12, 58–9, 77, 85, 88, 99, 117–18, 126, 128
  see also Human capacity and incapacity; Information processing
Autonomy 9, 88

Balance *see* Equilibrium; Symmetrical relations
Balancing operation 11, 16–17, 46–8, 55, 88, 97–8, 100, 102–103, 113, 121, 143, 151, 176, 194
  unbalanced 146
Barometer 19, 27, 41, 47, 110, 180, 208–209
Beats or intermittent tones 64, 75, 81, 113, 116, 118n., 128n., 143 & n., 144 & n., 148, 150–51, 154, 160, 195
  cause dissonance 53, 76, 113
  defined 53
Beauty *see* Excellence
Bells *see* Sound sources
Bodies 7–8, 15–16, 20, 40–41
  inelastic 16–17, 19, 45–6, 110, 140, 171, 198

Bodies, *cont.*
see also Elastic bodies; Human bodies; Material medium
Bowed instruments *see* Chordophones
Brain and nervous system 8, 16
auditory nerves 6–8, 58, 76, 124, 149
see also Auditory system
Break *see* Fracture
Breaking and varying tones 62–71, 91–2, 100–101, 147, 150–54, 172–73, 178, 195, 199
Breaking time *see* Tempo

Cadences *see* Musical closes
Capacity *see* Human capacity and incapacity
Castanets *see* Musical instruments
Cause 20, 37, 106
efficient 17, 45n., 139
material 7, 17
see also Impact
Centres 5, 45–6, 64, 66, 77, 85, 94, 97, 102, 106, 127, 129, 140, 145, 171–72, 177, 179
see also Equilibrium; Keys and keynotes; Vibration
Certainty and uncertainty 37, 108, 112, 123, 191
measure of uncertainty 119
Chordophones 65
bowed 65, 68
fretted and unfretted 79–80, 160
stringed instruments 51, 62n., 68
see also Musical instruments
Chords *see* Harmonic structures
Clocks and clockwork 70–75, 72, 86n., 94, 126, 184
watch spring 74, 208, 210n.
see also Sound sources
Closes *see* Musical closes
Coherence and incoherence *see* Musical coherence and incoherence
Coincidence theory of harmony 7, 9, 13, 52, 89–90, 97–9, 102–103, 112, 128, 141–44, 157, 160–61, 185n., 192–95
criticised 176–78
see also Consonance and dissonance
Collision *see* Impact
Colours *see* Light and colours
Computation *see* Account
Conception 14–15, 77
future 14–15, 77–8, 83–4, 86, 118–19, 124, 155–56, 161–63
past 14–15, 77–8, 86, 118, 124, 155–56, 164, 173–74, 184
present 14–15, 76–7, 84, 118, 124, 155, 161, 163–64, 174
Conjectures *see* Hypotheses
Consonance and dissonance 6, 50, 53–5, 73, 81, 103, 141, 143, 146, 160, 165, 176, 192
artificial discord 75
categorical order 13, 51–3, 55, 82, 91, 113, 142–43, 160
functional order 82–3, 104, 113, 153–55, 160–62
see also Beats; Coincidence theory of harmony; Harmony
Consort *see* Music
Constancy of period *see* Isochrony
Continuity 57–8, 86, 122n., 123, 149
Continuum 68
see also Mathematical continuum
Continuum hypotheses 1, 7–8, 17, 52, 105
kinematic variants 8, 10, 15, 20, 45, 48, 75, 90, 100, 107, 112
static variant 8, 15–16
Cords *see* Ropes
Correlations 13, 49, 60, 74, 92–3, 100, 112, 121, 145
Counting *see* Account
Coupled systems 57, 76, 126
Cycles 12n., 13, 49, 52, 85, 112
see also Rhythm; Vibration

Damped system 124
Damping 46, 56n., 57, 110, 122, 124, 149–50
device 150n.
Decorum *see* Excellence
Delight *see* Pleasure
Demonstrations 9n., 12, 74, 97
see also Experiments; Representations
Diagrams *see* Representations
Dissonance *see* Consonance and dissonance
Driving force *see* Impact
Drums *see* Musical instruments

Dulcimers *see* Musical instruments

Ear-brain system *see* Auditory system
Eardrum 3, 8, 12–13, 21, 76n., 111, 124, 137, 147, 179
Ears 1, 3, 5–7, 11, 14, 16, 41, 43, 45, 51, 53, 55, 58, 62n., 75–8, 80, 83–4, 96–9, 102–103, 111, 113, 124, 127–29nn., 139, 142, 149–50, 153, 155, 157, 160, 162–63, 171, 173, 176–78, 181, 191–92, 194
  musical 36, 76
  see also Auditory system; Hearing; Sense faculties; Sense organs
Echo 44n., 56n., 66, 150, 181–82
Edge tone systems 59, 113–15, 146n., 196
Elastic bodies 16, 45–8, 53, 56–7, 92n., 94, 101, 107, 110, 111n., 122n., 124n., 141, 144–45, 153, 171–73, 179–81, 198
  imaginary 202
  see also Material medium
Elastic force 7, 11, 16, 18–21, 40, 45–8, 53, 56, 60, 62, 65, 76, 94, 102, 107, 111, 116, 122n., 128, 140–41, 145–47, 150, 153, 171–73, 180–82, 193–94, 196, 198, 200
Elastic impact *see* Impact
Elastic medium *see* Material medium
Elastic process *see* Balancing operation; Impact
Elastic vibration *see* Vibration
Elasticity 10, 18–19, 40, 60, 87, 94, 121–22
  as motion against a resistance 16, 46–8, 55n., 68, 88, 103, 111, 113, 122, 139–40, 146, 151, 171–72
  criterion 46, 111, 180
  linear relation 92–94, 105, 107, 111, 113, 121–22, 127
  see also Impact; Resistance
Emission hypotheses 5, 7, 10n., 15–17
Emphasis *see* Rhythm
Energy 56–7, 94, 96, 98, 128, 141n., 150n.
  build-up 46, 53
  see also Heat; Light and colours; Sound
Environmental conditions 27, 41, 44–5, 96, 109–10, 139, 171, 179, 181, 182n., 190
Equalisation *see* Balancing operation
Equality of timing *see* Isochrony
Equilibrium 8n., 11, 66, 77–8, 88, 140n.
  position 46–7, 57, 63, 65, 94, 122, 128, 140, 145, 151, 172
  see also Centres
Êthos *see* Musical *êthos*
Excellence 1, 75–6, 81, 119, 160, 163, 186n., 187–88
  criterion 119, 190
  elegances 75, 83n. 162n.
  virtues 186, 187n.
Expectation *see* Conception
Experiences 2, 14, 37, 51n., 85, 100, 119, 124–25, 126n., 151, 178, 187n., 188, 190, 192n.
  see also Habituation; Memories; Musical practice
Experiments 2, 25n., 26, 30, 35, 37–9, 41, 48, 63, 70, 74, 89, 92, 96–7, 100, 109, 111–12, 122, 127, 135–36, 139, 140n., 175, 178, 182–83, 202
  candle flame 114–15, 196
  gun-blast timing 44–5, 121
  paper riders 69, 101
  toothed wheel 70–5, 126
  Torricellian 19, 41–2, 47n., 91–2, 95, 110, 139–40, 169–70, 171–72, 175, 178, 180
  vacuum 42–4, 95–6, 110, 121, 175–76
Eyes 39, 45, 53, 71n., 75, 118n., 142, 161, 178, 184–85, 191, 196

Faculty *see* Human capacity and incapacity; Mental faculties; Sense faculties
Fancy *see* Mental faculties
Fashion *see* Style
Flageolet tones 51, 62, 66, 68, 100, 138n., 151 & n., 152
Flageolets *see* Musical instruments
Flutes and flue pipes 58–60, 62, 92, 100, 113–15, 117, 146–47, 196–99
  whistles 60, 100, 114, 147, 152, 178
Force 11, 16, 45–6, 53, 57, 66, 68, 94, 102, 126n., 130, 142, 154, 170, 182, 191, 193, 198
  accelerating and retarding 46, 94,

Force, *cont.*
127–28, 198n.
vis impressa 198n.
see also Air or water resistance; Balancing operation; Damping; Elastic force; Friction; Mass
Formality *see* Musical formality
Fracture 48, 59, 140, 172
Frequencies *see* Infinite series
Frequency vii, 7, 13, 40, 53n., 57, 58n., 59n., 60, 62n., 78–9, 90, 99, 104, 115, 117, 128, 142n., 143n., 149n., 156n., 193n.
absolute 74–5
analysis 1, 126 & n.
and pitch 49–50, 71, 74, 112, 118, 121, 141, 144, 161, 172, 191
defined 2n., 49
numbers and ratios 52, 54–5, 74, 101, 104, 124, 126, 138, 142n.
see also Timing control; Timings; Vibration
Fretted and unfretted instruments *see* Chordophones
Friction 46, 110
Fundamental and secondary tones 51n., 53–5, 56n., 57, 65–6, 68, 70, 76, 82, 113, 129, 142–44, 150 & n.
natural or true tone 57, 145, 148
see also Harmonics and harmonic series; Flageolet tones; Overblown tones; Trumpet notes
Fundamental bass 54–5, 76, 82, 160–63
Future *see* Conception; Temporal framework and modalities

Grace notes 156, 165
Graphs *see* Representations
Gravity *see* Mass
Guns 30, 44–5, 47n., 140, 172, 182

Habit and Habituation 1, 3, 23, 33, 38, 78, 99, 125, 181, 183
see also Experiences; Musical practice
Harmonic motion *see* Motion
Harmonic progression 82n., 83
Harmonic relation 113, 129
see also Keys and keynotes; Fundamental bass
Harmonic structures 50, 51n., 54–5, 76, 77n., 82n, 83n., 82, 129
Harmonics and harmonic series 1–3, 40, 53–4, 55n., 62n., 66, 68n., 70n., 82, 83n., 91, 100–101, 113, 129
defined 2n.
see also Flageolet tones, Fundamental and secondary tones, Overblown tones
Harmony 66, 78, 103–104, 177–78
abstracted vs. composed 113, 195
colour 104
designed 75
see also Coincidence theory of harmony; Music; Musical formality
Harps *see* Musical instruments
Harpsichords *see* Musical instruments
Hautboys *see* Musical instruments
Hearing 4, 8, 14–15, 17, 51, 58, 76, 80, 85, 91, 117–18, 129n., 141, 143, 149n., 164, 174, 183–85, 192
theories 1, 3–4, 21, 76, 124n.
see also Audibility range; Auditory localisation
Heat 16, 18n., 94
thermal contraction or expansion 48n., 116
History 2, 4, 41, 87, 120n., 130, 165, 174, 186
Horns *see* Musical instruments
Human bodies 9, 14n., 117, 123, 182n., 184, 186, 190
Human capacity and incapacity 38, 51–2, 71n., 80, 99, 117, 139, 141, 181, 184–85, 187, 190–91
see also Audibility range; Mental faculties; Sense faculties
Humidity *see* Environmental conditions
Hypotheses 4, 15, 20, 26, 37–9, 75, 92, 102, 109, 111, 117–20, 123, 136, 139–40, 171–72, 178, 186, 194, 198n., 199, 209
hydrostatic 19–20, 26–7, 42, 209
see also Coincidence theory of harmony; Continuum hypotheses; Emission hypotheses; Impact
Hypothetical physics *see* Physics

Imagination *see* Mental faculties
Imitation *see* Musical *êthos*

Impact 7–8, 11, 16–7, 37n., 47, 59
as elastic process 11, 16, 20, 45–7, 57, 60, 92, 98, 107, 111–12, 122, 124n., 139–40, 145, 149, 153, 171–73, 180–83, 191–93, 198
see also Force
Incomparable *see* Inexpressible
Inertia *see* Mass
Inexpressible 75–6, 183, 186
Infinite 117, 161, 163, 180, 191
divisibility 49–50, 141, 178
see also Mathematical continuum
Infinitesimal 10, 98, 122n.
flux of a point 118
Information *see* Acoustic signals; Knowledge and ignorance
Information processing 39, 85, 117, 118n., 126, 127n.
conscious and subconscious 1, 3, 13, 54, 76–7, 85, 118, 126, 187n.
unconscious inference 125–6
Instruments *see* Aerophones; Chordophones; Measuring instruments; Musical instruments
Intellect *see* Mental faculties
Interference 40, 150n.
Intermittent tones *see* Beats
Involuntary *see* Voluntary and involuntary
Isochrony vii, 3, 45, 53, 63–4, 71–2, 94, 114, 122, 126–27, 192–94, 198
of small time intervals 6, 12, 14, 49, 66, 70–71, 107, 112, 128, 141, 144, 150, 172, 183 & n., 185, 191
of rhythm 85–6, 119, 126–27, 190
see also Pendula

Je ne sais quoi *see* Inexpressible
Jews-harps *see* Musical instruments
Just intonation 80, 81n., 101, 104, 129n., 144n.

Keyboard instruments 79–81
see also Musical instruments; Organs; Split keys
Keys and keynotes 76–9, 82–5, 113, 129, 155–58, 160–63, 165, 173–74
flat and sharp 81, 138n., 152, 156, 192
see also Musical formality; Tonality

Kinematics *see* Mechanics
Knowledge and ignorance 14, 38, 61, 95, 109, 123, 127n., 130, 136, 139, 99, 181, 187, 190
by acquaintance 109, 119, 134, 137, 179, 183, 186
foreknowledge 119, 187n.
information 73, 117–19
origins 75, 118
see also Conception; Experiences; Skill

Landscape 75, 186
Law 38, 39n., 108n.
moral 108n.
Laws 63, 92–4, 105, 107, 122–23, 127, 150, 202, 206
see also Correlations; Elasticity; Pendula; Principles
Light and colours 15, 17, 37, 41, 44–5, 70, 71n., 72, 75, 94, 96, 104, 106–107, 110, 139–40, 171, 176, 182, 184, 204
Life 186
better to be 105
Limitations *see* Human capacity and incapacity
Loudness *see* Rhythm; Sound intensity
Lutes *see* Musical instruments

Mass 57, 60 & n., 63n., 64, 128
gravity 27, 46, 65, 110, 130
weight 18–19, 60, 63–6, 112, 114, 148n., 151, 149, 182, 188, 193, 198
Mass-spring combination 60, 92, 115, 122
Material medium 10–12, 17, 32, 41–3, 45, 57, 96–7, 105–106, 110, 140, 171
deformable or elastic 1, 49, 94, 97, 110, 112, 122, 140, 171, 180–83
see also Air
Mathematical continuum 49–50, 117, 123–24, 139, 189
Mathematics 2–3, 10, 17, 30–31, 34, 37, 106, 108n., 202, 207, 208–11
see also Account
Matter *see* Bodies
Measure 119, 141n.
common 86, 98
linear 54, 60, 62–3, 65–6, 70, 68, 99–100, 140, 142–47, 170, 186, 188,

Measure, *cont.*
191
see also Account; Metre
Measuring instruments 45, 122
see also Monochords
Mechanical principles *see* Principles
Mechanical work 88, 97–8, 100, 111, 113, 181, 194
see also Balancing operation
Mechanics 3–4, 15, 21, 39, 46, 48, 55, 61–2, 76, 88, 103, 105n., 113, 121–22, 124n., 209
kinematics 10, 15, 106–107
Mechanisms 59, 88, 100, 123–24, 126
Medium *see* Material medium
Memories 102n., 125, 146n., 186n., 187, 190
see also Conception; Experiences; Knowledge
Memory 133–34
see also Mental faculties
Mental faculties 1, 3, 8–10, 15, 36n., 39–40, 56, 72–3, 76–8, 83–7, 99, 102–103, 108, 117–18, 125–26, 127n., 144, 155, 157, 161–65, 173–74, 184–86, 187n., 192
Metre 15, 21, 82, 85–6, 90, 126–27, 129, 161, 164–65, 173–74, 194
Mind *see* Attention process; Mental faculties
Models 11, 20, 46, 63–5, 87, 103n.
geometrical 9
physical vii, 4, 17, 21, 47–8, 86, 88, 122–24, 126, 198
see also Pendula
Modes 88, 166n.
see also Keys and keynotes; Style
Modes of time *see* Metre
Modes of vibration *see* Vibration
Modulation *see* Keys and keynotes
Monochords 51, 79, 141, 157
division 51–2, 54, 80, 104, 141, 150, 152, 156, 165–68
see also Scales and scale division
Motion 10–11, 15–16, 20, 28, 49, 53, 97, 105, 196, 202
muscular 27
parallel 83–4, 161
random 18, 47
simple harmonic 40, 53, 94, 105, 112n., 125, 128n.
spontaneous 50
sympathetic 8, 70–71, 194
vortical 7, 59, 113–15, 146–47, 182n., 196
see also Elasticity; Vibration; Wave
Muscles *see* Motion
Music 39, 75, 109n., 139, 164, 183, 195, 202
as a form of communication 119
as movement in time 1, 85, 124
in ancient time 78, 88, 153, 166, 173–74
in one part 76–8, 155–56, 162, 164, 173, 178
in several parts 14, 21, 81–5, 100, 127, 160–64, 173–74, 178, 194
Musical air 15, 77, 83n., 85
Musical basis or foundation *see* Fundamental bass
Musical coherence and incoherence 77, 84, 119, 129–30, 155–56
see also Musical formality; Tonality
Musical closes 77–8, 83, 138, 155–56, 161–62
Musical *êthos* 15, 87–8, 163, 165–66, 173–74
Musical formality 77–8, 82, 84, 86, 88, 156, 160, 162–64, 173
see also Musical coherence and incoherence
Musical instruments 39–40, 44, 51n., 56–8, 75–6, 79–81, 100, 126, 128–29, 135, 141–42, 144, 149–50, 156, 158, 160, 162n., 172, 178, 195
castanets 85, 164
drums 85, 87, 143, 148n., 164–65, 192, 195
dulcimers 150, 172
flageolets 60, 147n., 196
harps 160
harpsichords 32n., 149–50, 160, 172, 192
hautboys 146
horns 58, 165
jews-harps 148
lutes 58, 160
recorders 60, 147n.
sackbuts 148–49, 152–53, 173

Musical instruments, *cont.*
  shawms 148
  trumpets 58, 62, 66, 68, 101, 102n., 148–49, 152–53, 173
  trumpets marine 66–8, 101, 152, 173
  violins 25n., 57, 79, 149, 160, 172
  viols 23n., 51n., 65, 160, 192
  see also Aerophones; Chordophones; Keyboard instruments; Organs; Sound sources; Wind instruments
Musical intervals *see* Coincidence theory of harmony; Keys and keynotes; Octave identity; Scales and scale division
Musical meanings *see* Musical semantics
Musical metre *see* Metre
Musical practice 1–2, 4, 33, 62n., 76, 79, 81, 83n., 183
Musical rhythm *see* Rhythm
Musical science 2, 4, 21, 124
Musical semantics 39, 76–8, 82, 83n., 88, 119, 124
  see also Musical coherence and incoherence; Musical *êthos*
Musical time system 86
  see also Metre; Rhythm; Tempo
Mysticism 51

Nature 42n., 77n., 82, 88, 108, 160, 166, 174, 184, 187
Nerves *see* Brain and nervous system
Nodal forms and phenomena 69, 101
  see also Breaking and varying tones
Node 3, 40, 68–9, 101, 150
  antinode 115
Noise 10, 14n., 33, 48, 50, 55, 75, 117, 127, 129, 140n., 141, 175, 181, 185, 190
  crack 48, 140, 172, 180
Norms 52, 55, 82, 84, 102, 126–27
Note 40, 76, 155
  measure or time note 86, 165
  notation 83n., 86
  see also Grace notes; Keys and keynotes

Octave identity 55, 82, 142, 144, 161
Organs 33, 51n., 59–61, 79n., 81n., 115–16, 143, 160, 192, 195
  stops 51n., 58, 147, 148n.
  tuning 116, 144, 147, 160, 199
  voicing 116
  see also Flutes and flue pipes; Pipes
Oscillating systems *see* Vibrating systems
Oscillation *see* Vibration
Oscillators *see* Vibrators
Overblowing 62, 149n., 152n.
Overblown tones 51, 62, 68, 100, 152n.
  see also Trumpet notes
Overtones 73
  see also Harmonics and harmonic series

Pain 21, 55, 182, 184–85, 188, 190–91
Partial tones *see* Harmonics and harmonic series
Passions *see* Affect; Experiences; Pain; Pleasure
Past *see* Conception; Temporal framework and modalities
Pattern 7, 48, 54, 56, 57n., 58n., 73, 101, 125, 127
  recognition 13, 83n., 85, 118n.
  see also Periodicity; Symmetrical relations
Pendula 3, 46, 49, 53n., 63–6, 86, 90, 92n., 94, 98, 105, 112, 114, 122–26, 150–51, 173, 192, 198, 208
  laws 63, 86, 122, 127, 150
  see also Motion; Vibration
Perceptible and imperceptible *see* Information processing
Perception theories 1, 3, 9
  see also Perceptual cognition; Sensory perception
Perceptual cognition 1, 9, 39, 77–8, 85, 99
  see also Information processing
Percussion *see* Impact
Perfection *see* Excellence
Periodicity 13, 49, 54, 73, 77, 85, 90, 118–19, 124, 128
  see also Frequency; Rhythm
Phase vii, 40, 98–9, 102–103, 105, 112n., 113, 121, 124, 126n., 128, 178
  see also Synchrony
Philosophies 2, 5, 8, 10–11, 15, 20, 23, 32, 34–5, 38, 44, 46–7, 90, 94n., 107n., 121, 127, 201, 203, 210

Philosophies, *cont.*
empirical 1, 3, 14, 78, 80, 210
rational 52, 80
Philosophising 37–8, 105, 107–108, 119n., 123, 210
Philosophy of music 1–4, 32, 36n., 39n., 40, 89, 90n., 105, 107, 109, 115n., 121, 124, 127, 130, 201
Physics 16, 11n., 49n., 96, 109n., 202
hydrostatics 8n., 26
hypothetical 4, 5–21, 121
see also Acoustics
Pipes 32–3, 56, 59, 60–61, 68, 70–71, 94, 100 & n., 116, 123, 145–48, 150, 153–54, 160, 172, 191, 195, 199–200
lip-vibrated 58, 60, 62, 148–49
reed 58, 60, 148, 196
theories 59n., 128n.
see also Aerophones; Air columns; Flutes and flue pipes; Musical instruments; Organs
Pitch 21, 75, 85, 124, 126–27, 129, 139n., 191n.
see also Frequency; Tone
Pleasantness and unpleasantness *see* Consonance and dissonance
Pleasure 1, 21, 33, 51, 72, 85, 88, 162, 164, 183, 192, 194
source of 14–15, 113, 119, 127, 185–87
Present *see* Conception; Temporal framework and modalities
Pressure or expansive force *see* Elastic force
Principles 1, 3, 18n., 38, 88, 109, 125, 183, 188
a priori 37–8, 52
epistemological 14
mechanical 10–11, 53–5, 62, 82, 92, 96, 100, 106, 107n., 112–14, 193n., 198
see also Correlations; Musical formality
Proportionable times *see* Periodicity; Rhythm
Proportions of time *see* Metre
Pulse 10, 40
beat 45
front 49, 52, 70, 98, 103
see also Vibration; Wave
Pulse doctrine *see* Coincidence theory of harmony

Quantity *see* Rhythm

Reasoning 40
see also Philosophising
Reckoning *see* Account
Recorders *see* Musical instruments
Regard *see* Mental faculties
Relative timing *see* Phase
Remembrance *see* Conception
Representations 7, 9, 37
geometric 12
graphic 54–6, 70, 73, 89, 123–24, 126
numeric 54, 124, 138
sonogram 56
see also Analogues; Models
Resistance 10–11, 111
defined 11n.
see also Air or water resistance; Elasticity; Friction
Resonance 46, 51n., 53, 55, 57, 58n., 75, 98, 100–101, 103–104, 115, 124, 142, 150n., 153, 173, 176n.
auditory 1, 3, 76, 124
see also Motion; Sound reinforcement and destruction; Strings; Synchrony
Resonators 57–8, 65n., 66, 76, 98–100, 114, 148n., 149–50
selective 153, 173
Rest *see* Equilibrium
Restoring force *see* Elastic force
Retarding force *see* Force
Rhythm 85–6, 90, 124, 126, 129, 164
accent ('emphasis') 15, 77–8, 85–6, 156–57, 165, 173
quantity ('length') 78, 156, 165, 173
stress ('loudness') 21, 78, 155–56, 165, 173
see also Periodicity
Ropes 74, 112, 182, 191

Sackbuts *see* Musical instruments
Satisfaction *see* Pleasure
Scales and scale division 31, 52n., 78–80, 91, 101, 104, 117, 129, 141, 144, 150, 153, 157–60, 172–73, 178, 192
schisms 78–81, 144, 156–58, 160, 173

Scales and scale division, *cont.*
see also Monochords
Scepticism 108
Schisms *see* Scales and scale division
Sciences 10, 46n., 94, 105, 121–23
see also Acoustics; Elasticity; Mechanics; Musical science
Secondary tones *see* Fundamental and secondary tones
Segmental vibration *see* Breaking and varying tones
Semantics of music *see* Musical semantics
Senario 51, 54
Sensation *see* Conception; Sensory perception
Sense 133–34
see also Ears; Hearing; Sense faculties
Sense faculties 11, 52, 56, 72, 102n., 117–18, 129n., 139, 171, 181, 184–88, 190–94, 162, 169, 171
intuition 38, 125
taste 1, 113, 194–95
see also Ears; Eyes; Hearing
Sense impressions *see* Sensory perception
Sense organs 1, 41, 72, 103, 139–40, 143, 171
see also Brain and nervous system; Eardrum; Ears
Senses *see* Sense faculties
Sensible and insensible *see* Information processing
Sensory perception 6, 8–9, 13–14, 39, 49, 51–2, 55, 72, 76–7, 82, 85, 102–103, 115, 117–18, 125–26, 149, 155, 181, 184, 186, 190–91
see also Information processing
Shawms *see* Musical instruments
Shock *see* Impact
Signs 1, 38, 40, 125, 129
Simulations *see* Models
Skill 15, 33, 120, 156, 164, 175, 178, 190, 203
Smoothness and roughness *see* Consonance and dissonance
Sound 74–5, 94, 121
defined 5, 8, 11, 48–9, 70, 94, 122, 182
see also Echo; Noise; Tone
Sound augmentation *see* Sound reinforcement and destruction
Soundboards *see* Resonators
Sound generation 1, 3, 45, 47, 49, 57, 70, 106, 111–12, 140, 145, 153, 171–72, 179, 181–83
see also Flutes and flue pipes; Impact; Pipes
Sound generators *see* Sound sources
Sound intensity 41–3, 53, 82n., 139, 147, 171, 175–76, 179, 181, 185
see also Sound reinforcement and destruction
Sound propagation 1, 32, 176
distance 5, 41, 44, 139–40, 171, 179, 181
instantaneous 9, 46
in time 10, 41, 46, 75, 103, 106, 139, 171, 176, 182n.
path traversed 8, 17, 41–2, 43, 96–7, 106, 112–13, 139, 171, 176–76, 182
speed 41, 44–5, 48, 97, 100, 107, 110, 139, 175, 179–80
see also Analogues; Continuum hypotheses; Emission hypotheses; Environmental conditions; Material medium; Wave propagation
Sound reception 1, 177
see also Ears; Hearing
Sound reinforcement and destruction 53, 57, 66, 75, 99–100, 102, 104, 142, 149, 152–54, 173, 178, 193–94, 198n.
see also Resonance; Resonators
Sound sources 1, 3, 5, 7, 11–12, 42n., 46, 49, 53, 56–7, 70, 91–2, 95–6, 97–8, 100, 103, 105–106, 110–12, 127, 136–39, 141, 143, 145–49, 151, 169–71, 173, 176–78, 179–80, 183, 183, 190, 193–94
bells 7–9, 10, 42n., 43, 56, 57n., 92, 94–5, 110, 140, 144, 150, 165n., 172, 175–76, 191
watches 42–3, 175–76
toothed wheels 70–75, 126
see also Musical instruments; Organs
Sound wave *see* Wave
Space 8, 10, 49, 99, 117–18, 190–91
Speech sounds *see* Voice
Split keys 80, 158
Spring *see* Elastic force; Vibrators

Springs 48, 107, 123, 198
helical, 92, 111, 123, 124n., 180n., 182
see also Clocks and clockwork; Mass-spring combination
Stress *see* Rhythm
Stringed instruments *see* Chordophones; Musical instruments
Strings 3, 6–7, 9, 12–14, 16–17, 44, 46, 50, 52–5, 57, 58n., 65–9, 70–71, 90, 92, 94, 97–8, 102, 104, 112, 115, 123–24, 142–44, 149, 152, 156, 158, 172, 180n., 185, 191–95
equally, unequally stretched 6, 12, 14, 44, 97, 100–101, 103–104, 112, 142, 149, 153, 173, 176, 193–94
stiffness factor 62n.
sympathetic 51n., 66, 70, 142
true and false 62–5, 150–51, 191
see also Monochords; Ropes; Vibrators
Style 32, 33, 87, 101n., 108, 135, 156, 165, 174
fashion or mode 156, 165, 174
Symbols *see* Signs
Symmetrical relations 54, 82, 84, 129, 138, 143, 187, 190
Symptoms *see* Signs
Synchrony 3, 13n., 53–5, 57–9, 62, 64–6, 68–70, 97–8, 100–102, 124, 151 & n., 154 & n., 176, 183n.
in the same measure, rate or time 3, 12–13, 21, 55, 64, 66, 83, 102–103, 144, 149, 152–53, 161, 173, 177, 193
keeping time with 21, 27, 154n.
see also Coincidence theory of harmony; Resonance; Timings; Vibration
Systems 59n., 136, 153n.
see also Auditory system; Coupled systems; Edge tone systems; Tunings; Vibrating systems

Taste *see* Sense faculties
Temperaments 79–81, 91, 128–29, 144n., 156n., 158–60, 167–70
Temperature *see* Environmental conditions
Tempo 75, 85–6, 129, 164–65, 173–74, 194
breaking time 119, 127, 190
see also Metre; Rhythm
Temporal framework and modalities 14, 77, 87, 123–24, 164, 174
before/after 75, 77, 86, 118, 155–56, 162–64, 173
Temporal phenomena and processes vii, 1, 12n., 56
Tension or stretching force *see* Elastic force
Timbre *see* Tone quality
Time 1, 6, 8–10, 13, 56, 74, 117–19, 122–24, 127, 191
see also Musical time system; Temporal framework and modalities
Time as experienced *see* Attention process; Conception
Time-keeping methods 45, 86–7, 165n.
see also Account; Experiments
Timing control 21, 53–4, 56–7, 70, 102, 113–14, 144–45, 150, 193, 198–200
see also Elastic force
Timings vii, 12, 14, 21, 70, 76, 98
see also Frequency; Isochrony; Phase; Synchrony
Tonality 1–2, 4, 40, 77–8, 129–30
see also Keys and keynotes; Musical formality
Tone 40, 50, 63, 75, 117, 127, 170
as compound 54, 70, 51n., 54, 82n., 118, 125–26, 138n., 184, 194
defined 49, 50, 54, 70–71, 75, 92, 101, 112, 139, 149, 183–84, 189
quality 70n., 99, 191
see also Beats; Frequency; Pitch
Tones *see* Breaking and varying tones; Flageolet tones; Fundamental and secondary tones; Overblown tones; Scales and scale division
Torricellian vacuum *see* Experiments
Trumpet notes 62, 101, 148n., 149, 152
see also Overblown tones
Trumpets *see* Musical instruments
Trumpets marine *see* Musical instruments
Truth 9, 14, 37–8, 41, 108, 109n., 183
Tuning 79, 81, 116, 157–58, 160
by beats 116
see also Organs
Tunings 51n., 129n.
see also Just Intonation; Temperaments

Uncertainty *see* Certainty and uncertainty
Understanding *see* Mental faculties
Undulation 3, 12n., 17, 97, 177

Vibrating segments *see* Breaking and varying tones
Vibrating systems 55n., 57, 60, 123–25, 127–28
  see also Coupled systems; Mass-spring combinations; Pendula
Vibration 3–4, 10, 40, 56, 62, 94, 123, 142, 145–46, 149–51, 169–70, 190
  amplitude 46, 53, 55n., 57, 59n., 63, 107n., 115, 126, 128, 150n.
  free and forced 57
  modes 3, 53–4, 62, 68, 70, 76, 101, 102n., 115, 123–24
  pendular 3, 45–6, 49, 53–4, 63–6, 71, 75, 77, 85–6, 90, 94, 98, 101, 105, 107, 112, 114, 117, 122–26, 128n., 140, 144–45, 151, 190, 194, 196
  see also Frequency; Isochrony; Synchrony
Vibrators 60, 128
  acoustic 57, 58, 60, 76, 100, 147–48
Violins *see* Musical instruments
Viols *see* Musical instruments
Virtue *see* Excellence
Voice 23n., 32n., 79, 153, 160, 162n., 173
  speech and speech sounds 71–2, 88
Voluntary and involuntary 27, 187n.

Water 12, 17, 96–7, 106, 109, 114–15, 145, 177, 179, 184, 198
Water resistance *see* Air or water resistance
Wave 40, 58, 98, 106, 107n., 109–10, 117, 145, 193
  front 96–8, 106
  see also Undulation
Wave propagation 1, 2n., 7, 10, 12, 94, 96, 122, 181–82, 193
  longitudinal 106, 111–12, 123
  transverse 48, 106, 111–12
Wave surfaces 11–12, 17, 97–8, 106, 109 110–11, 177, 179, 181, 182n., 193
Whips 140, 172
Whistles *see* Flutes and flue pipes
Whistling 147, 182n.
  see also Flageolet tones
Wind *see* Environmental conditions
Wind instruments *see* Aerophones